Romanian Agriculture and Transition toward the EU

Rural Economies in Transition

Series Editors

Much of the world's rural population lives in transition countries that are attempting to transform their economies from the centrally planned socialist system to a market-oriented system. The Rural Economies in Transition series evaluates the deep impact these transformations are having in Central and Eastern Europe, the former Soviet Union, and Asia. Based on original empirical work by international teams at the cutting edge of research, each volume presents the results of one complete empirical study devoted to a single country or region. The editors have designed the series to provide agricultural, development, and political economists, as well as scholars in other disciplines, with authoritative and up-to-date factual information about rural transition. The editors welcome manuscript submissions from scholars engaged in original empirical research in the field.

Titles in the Series

Private Agriculture in Armenia, by Zvi Lerman and Astghik Mirzakhanian
Transition, Institutions, and the Rural Sector, edited by Max Spoor
Romanian Agriculture and Transition toward the EU, edited by
Sophia Davidova and Kenneth J. Thomson

Romanian Agriculture and Transition toward the EU

Edited by Sophia Davidova
and Kenneth J. Thomson

LEXINGTON BOOKS
Lanham • Boulder • New York • Oxford

LEXINGTON BOOKS

Published in the United States of America
by Lexington Books
A Member of the Rowman & Littlefield Publishing Group
4720 Boston Way, Lanham, Maryland 20706

PO Box 317
Oxford
OX2 9RU, UK

British Library Cataloguing in Publication Information Available

Library of Congress Control Number: 2002115567

ISBN 978-0-7391-0518-4

Printed in the United States of America

™ The paper used in this publication meets the minimum requirements of American National Standard for Information Sciences—Permanence of Paper for Printed Library Materials, ANSI/NISO Z39.48–1992.

Contents

Figures

Tables

Preface

Romania is one of the largest, yet one of the poorest, of the ten Central and Eastern European countries (CEECs) which have applied for membership of the European Union (EU). Though much smaller than in the mid-1990s, its agriculture is still a major component of the national economy and is even more important in the social life and environment of the country. As the EU enlargement negotiations in agriculture with Romania are still to come, the variety of problems to be faced by agricultural policy-makers in both Brussels and Bucharest needs to be fully explored and appreciated.

The main aim of this book is to report the results of a number of recent studies on different aspects of the reform policies in the Romanian agri-food sector during its transition toward a market-oriented system. Most are based on the EU Pre-Accession Financial Programme (PHARE) project RO9804.03.01 carried out during 1998–1999 by the Agricultural Reform and EU Alignment (AREA) Group in the Ministry of Agriculture and Food (MoAF), Bucharest. However, the chapters have been modified for this publication and should not be taken as representing the views of the Government of Romania, the European Commission, or the other institutions for which the authors worked at the time or subsequently.

Romania poses transition and EU accession problems in particularly acute form. However, the conclusions reached are not only of domestic importance but are also relevant to several other economies in Central and Eastern Europe and indeed for many post-socialist countries where the agri-food sector still makes a major contribution to the economy and is an important vehicle for rural development. In addition, agricultural questions are proving to be some of the most difficult to resolve in the EU enlargement negotiations. The nature of these difficulties, before and after accessions, deserves full consideration.

The first half of this book is concerned with the problems in Romanian agriculture in the transition period. Chapter 1 sets the scene by providing an overview of the macroeconomic context, the place of agriculture in the Romanian economy, and the potential of the agri-food sector. Chapter 2 discusses Romanian agricultural and trade reforms and the need for sustaining the reform path in order to increase the efficiency and competitiveness of agriculture. Chapter 3 focuses on the factors that affect the development of individual farming in Romania and the policies that could remove the impediments to such development. Chapter 4 uses survey results to assess the success of the rural financial policy for the agri-food sector, concluding that both the supply of and demand for rural credit for small individual farmers needs to be improved and suggesting ways in which this could be achieved. Chapter 5 extends the analysis of credit issues by examining the sources of constrained access to credit for emerging commercial farming. In chapter 6, a tool is developed for low-cost and rapid means of appraising the financial situation of farm households and businesses and for investigating the effects of different policy measures. Chapter 7 uses a farm level model of risk to assess whether Romania should introduce a price stabilization scheme for wheat and concludes that this is not an urgent priority for the improvement of farm level decision making in that country.

The second half of the book turns to the relationship between Romanian agriculture and the EU. The international competitiveness of Romanian agriculture is assessed in chapter 8, while chapter 9 discusses the impact of the pre-accession Europe Agreement on agricultural trade between Romania and the EU, concluding that it has been insignificant and suggesting reasons for this. Chapter 10 reviews the ways in which the Special Accession Programme for Agriculture and Rural Development (SAPARD) can help in rural development in Romania and the need for flexibility and simplicity in its applications. In chapter 11, a quantitative economic model is applied to estimate some of the major economic effects related to Romanian accession to the EU. The same model is used in chapter 12 in the examination of Romania's World Trade Organization commitments both before and upon accession to the EU. The conclusion is reached that these commitments are unlikely to cause problems on accession but may do so beforehand in the areas on domestic support and export subsidies.

A final chapter highlights the major findings of these wide-ranging studies, identifies some gaps and uncertainties, and makes some suggestions about the future direction for Romanian agriculture and how this direction might be analyzed.

If the declared intention of achieving accession to the EU by the year 2007 is to be realized, Romania now faces some critical decisions. The lessons of the last few years, both in the country and in other applicant states, need to be understood and applied if more time is not to be lost and the enthusiasm of the first post socialist years is not to evaporate. Reform and progress in Romania's agricultural sector would simultaneously provide alternative employment for at least some of the rural work force, reduce food costs to the poorer sections of the national population and to the state budget, and improve Romania's export performance by providing higher-value processed food and drink products. In doing this, accession to the EU would become easier and quicker, with benefits to both sides. A large and economically stable free trade area would reduce the problems of fluctuating trade quantities and prices and make it easier to cope with farm crises that can always be expected.

It is easy to point out the problems of Romanian agriculture and not so difficult to recommend solutions. Carrying them out in the real world of policy design and implementation is the task of all those carrying responsibilities in industry and/or government. If this book can assist in the huge work of education, enlightenment, and persuasion that lies ahead, its preparation will have been worthwhile.

Acknowledgments

A large number of people and institutions need to be thanked for providing the "infrastructure" necessary for the production of this book. Among these, Professor Stelian Dinu Gavrilescu, Minister of Agriculture and Food during the lifetime of the project and Director of the Institute of Agrarian Economics at the Academy of Sciences in Bucharest, gave unstinted advice, information, and support throughout the project. Dr. Stelian Minoiu, head of European Integration, Cooperation and International Relations at the MoAF at that time, chaired the Project Steering Committee, with Mrs. Roxana Ionescu, Director of the EU PHARE Policy Management Unit, as cochairman, both were very supportive and open-minded. The general secretary of the MoAF, Mrs. Cristina Esanu, also provided support and useful comments.

Other Romanians under training played an important part in the work reported in this book and benefited directly or indirectly from financial support from the EU. Most are named in the relevant chapters but they deserve recognition here. They include (in alphabetical order): Silviu Balanica, Simona Bara, Victoria Burtea, Virginia Campeanu, Carmen Firici, Florica Grigore, Jacqueline Leonte, Iulia Carciumaru, Simona Lubieniechi, Simona Popov, and Mihaela Preda. Among those from outside Romania, we thank Erik Mathijs, Hamish Gow, Alexander Sarris, and, in particular, Dr. Simon Davies of the Scottish Agricultural College, advisor to the PHARE project throughout, and Professor Zvi Lerman at the Hebrew University of Jerusalem, who provided most useful commentary on chapter drafts. Finally, at Imperial College Wye, Helen Bright acted as our eagle-eyed, efficient, and understanding editorial advisor.

Sophia Davidova and Ken Thomson
Imperial College Wye (University of London), and University of Aberdeen
June 2002

Abbreviations

ACE	Assistance for Cooperation in Economics
AS	Agricultural Society
AMS	Aggregate Measure of Support
ANPA	National Agency for Agricultural Products
ASAL	Agricultural Structural Adjustment Loan
BA	Banca Agricolă
BIR	Banca Internationala a Religiilor
BCR	Banca Comerciala Romana
BRD	Romanian Development Bank
CAP	Common Agricultural Policy
CC	Cooperative de Credit
CCPC	Commercial Company with predominantly Private Capital
CCSC	Commercial Company with predominantly State Capital
CEC	Casa de Economii şi Consemnaţiuri
CEEC	Central and Eastern European Country
CEFTA	Central European Free Trade Agreement
c.i.f.	cost, insurance, freight
CIS	Commonwealth of Independent States
CN	Combined Nomenclature (for EU trade statistics)
CPI	Consumer Price Index
CSE	Consumer Support Estimate
DGAgri	Directorate General for Agriculture
DRC	Domestic Resource Cost
EA	Europe Agreement or Association Agreement
EAGGF	European Agricultural Guidance and Guarantee Fund
EBRD	European Bank of Reconstruction and Development
EC	European Commission
ECU	European Currency Unit

EPR	Effective Protection Rate
EU	European Union
FA	Farmers' Association
FADN	Farm Accountancy Data Network
FAO	Food and Agriculture Organization
FDI	Foreign Direct Investment
f.o.b.	free on board
GATT	General Agreement on Tariffs and Trade
GDP	Gross Domestic Product
HS	Harmonized Customs System
IMF	International Monetary Fund
ISPA	Instrument for Structural Policies for Pre-Accession
MFN	Most Favored Nation
MIT	Ministry of Industry and Trade (in Romania)
MoAF	Ministry of Agriculture and Food (in Romania)
NBR	National Bank of Romania
NPAA	National Program for the Adoption of the Acquis
NPC	Nominal Protection Coefficient
OECD	Organization for Economic Cooperation and Development
PAM	Policy Analysis Matrix
PHARE	EU Pre-Accession Financial Programme
PSE	Producer Support Estimate
SAPARD	Special Accession Programme for Agriculture and Rural Development
SPP	Special Preparatory Program
TAL	Total Agricultural Land
TSE	Total Support Estimate
URAA	Uruguay Round Agreement on Agriculture
U.S.(A)	United States (of America)
WTO	World Trade Organization
ha	hectare(s)
hl	hectoliter(s)
km	kilometer(s)
ppm	parts per million
t	metric tonne(s)

As data in many tables have been rounded, minor discrepancies may appear between totals and their parts.

Chapter 1

Introduction: The Context for Reform and Accession

Kenneth J. Thomson

The Macroeconomic Context

In terms of land and population, Romania is the second largest (after Poland) of the ten Central and Eastern European countries (CEECs) which have applied for membership of the European Union (EU). However, it is also one of the poorest and most backward of these countries, with income per head at $5,900 compared to Bulgaria at $6,200 and Poland at $8,500 (purchasing power parity, 2000 estimates).

In 1989, after decades of strict state control, the Romanian macroeconomy was more balanced than that of most of the other CEECs. Annual inflation was only 0.9 percent, the current account was positive at $2.3 billion, and the state budget was in surplus (Zecchini 1997, 1–34). Therefore, following the overthrow of the dictator Ceaucescu in December 1989, macroeconomic imbalances did not prompt radical policy changes. Moreover, there was no organized alternative party or movement in existence to take over the reins of power and to determine whether the transition toward a market-oriented economy should begin with a relatively short period of "shock therapy" as in Poland or in a more gradual way.

Instead, a period of slow and uncertain political and economic reform ensued under a coalition government made up mostly of officials in the former regime. State involvement in the economy remained substantial, while price and trade liberalization proceeded piecemeal over several years. A combination of weakened state power, some uncontrolled private activity, and loss of external markets resulted in massive underinvestment, a demoralized and unskilled labor force, and a negative external image.

In November 1996, a new center-right government came in with sufficient confidence to make a coherent start on programs of fiscal and monetary policy reform and on some restructuring activities such as more or less complete trade liberalization, the privatization of state enterprises, and land reform. As a result of government commitments to reform and pressure from the international financial institutions, price and trade policies were largely liberalized in 1997. Nevertheless, macroeconomic stabilization could not be rigorously imposed, and 1997, 1998, and 1999 were again difficult years, with the Gross Domestic Product (GDP) falling once again, and inflation (following removal of some price controls) above 50 percent.

Figure 1.1 shows the development of Romanian GDP and inflation over the decade as a whole. GDP fell by over a third between 1989 and 1992, with a partial recovery to 1996, while annual inflation exceeded 100 percent per year between 1991 and 1994, before falling to around 40 percent per year in 1995 and 1996. Strong negative correlation between these two variables is obvious, along with generally poor GDP performance and consumer inflation seldom below 50 percent.

Lack of economic development over this period led to lack of employment, although reluctance to allow major enterprises to undergo bankruptcy and a certain amount of "return to the land," held down recorded levels of underemployment to single figures. Nevertheless, poverty became endemic, creating both social difficulties (e.g., the unpopularity of Roma and Germanic groups) and policy problems (e.g., resistance to dismantling price controls). Low birth rates, out-migration, and high crime rates became major social concerns. Throughout the 1990s, the balance of trade, current account, and governmental budget balance were all negative, resulting in a fast-depreciating (but persistently overvalued) national currency (the lei). However, in the year 2000, the Romanian economy began to show some signs of stabilization and recovery after years of fragility, with a significant rise in GDP, improved exports, and stable, though still very high, inflation.

Figure 1.1 GDP Change and Inflation, Romania, 1989–2000

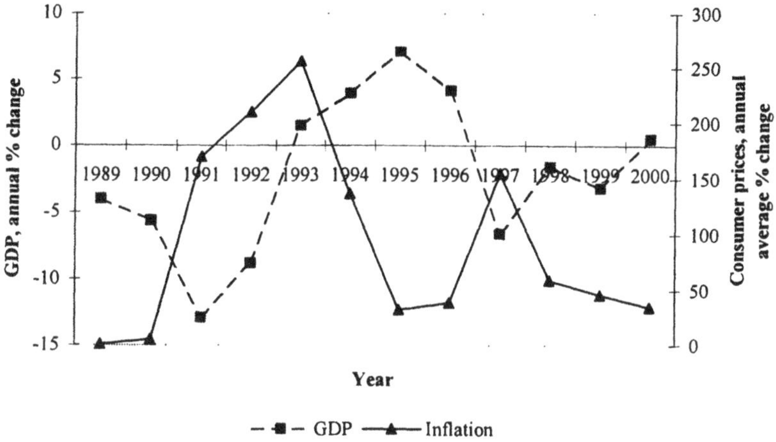

Source: EBRD (1998, 2000)

Table 1.1 compares Romania with other CEECs in terms of various European Bank of Reconstruction and Development (EBRD) indicators of progress in transition by the year 2000. On a scale of one to four (the latter representing the standards and performance of advanced industrial economies), Romania has a high score for the liberalization of its trade and foreign exchange system, indicating full current-account convertibility and virtual withdrawal of state ministries and state trading companies from exports and imports (for nonagricultural goods only). However, all other scores are lower, with the lowest (two) indicating lack of progress in restructuring governance and enterprise, competition policy, and the development of financial institutions. An overall score (treating + and — as 0.25 higher and lower than the corresponding integer) leaves Romania with the lowest total among all ten CEECs, just behind Bulgaria. Private sector indicators, given on the right of table 1.1, reflect this performance, and may explain part of it, since once again Romania is at or near the bottom ranking, with only 60 percent of the economy in private hands and very low inflows of foreign direct investment (FDI) over the period 1989–1997.

Table 1.1 Romania's Progress in Transition in Comparison with Other EU Applicants

Country	EBRD Scores											
	Enterprises			Markets and Trade			Financial Institutions		Legal Transition	Average EBRD score	Private sector in GDP, mid-1999 (%)	Cumulative per capita FDI inflows, 1989–1997 ($)
	Large-scale privatization	Small-scale privatization	Governance and enterprise restructuring	Price liberalization	Trade and foreign exchange system	Competition policy	Banking reform and interest rate liberalization	Security markets and nonbank financial institutions	Commercial law			
Romania	**3–**	**4–**	**2**	**3**	**4**	**2**	**3–**	**2**	**3+**	**2.83**	**60**	**149**
Bulgaria	3	3+	2+	3	4+	2	3–	2	4–	2.92	60	147
Czech Republic	4	4+	3	3	4+	3	3+	3	3	3.42	80	823
Estonia	4	4+	3	3	4	3–	4–	3	4–	3.50	75	695
Hungary	4	4+	3+	3+	4+	3	4	3+	4–	3.67	80	1,667
Latvia	3	4	3–	3	4+	3–	3	2+	3	3.11	65	543
Lithuania	3	4+	3–	3	4	2+	3	3–	3+	3.14	70	344
Poland	3+	4+	3	3+	4+	3	3+	3+	3+	3.42	65	321
Slovakia	4	4+	3	3	4+	3	3–	2+	3	3.28	75	227
Slovenia	3+	4+	3–	3	4+	2	3+	3	4	3.31	55	639

Source: EBRD (1998, 1999)

Agriculture in the Romanian Economy

Agriculture remains central to the Romanian economy, but the background of slow macroeconomic reform and persistent recession has resulted in unfavorable conditions for agricultural development. Indeed, farming has both contributed to, and suffered from, the macroeconomic sluggishness and instability. Although agriculture's share of GDP has fallen from over 20 percent in 1995 to around 11 percent in 2001, it accounts for around 40 percent of employment, a figure that has risen during the 1990s.

Before transition began, collective and state farms occupied nearly 90 percent of the agricultural land in the country. The dismantling of collective farms—theoretically, voluntary farmer co-operatives, but in practice totally state dominated—began in 1990, through a process of land restitution subject to upper area limits. By 1998, 84 percent of the land due for privatization under the 1991 Land Law had been transferred to 3.8 million persons.

The redistribution of farmland has led to an extremely fragmented ownership structure of about 4 million holdings with an average size of less than 3 ha. However, actual farming practice is much less fragmented, due to the formation of about 3,500 agricultural societies (ASs) and about 6,000 farmers' associations (FAs) with average sizes in 1999 of 400 ha and 140 ha, respectively. ASs are legally registered entities similar to producer cooperatives. FAs are spontaneous arrangements between farmers to farm in association and do not have a legal status. ASs account for about 10 percent and FAs for 6 percent of the total agricultural area, respectively, and the public sector (ex-state farms, referred to as commercial companies with predominantly state capital [CCSCs], and other state land holdings), about 15 percent. Nevertheless, most of the land is now farmed by individual independent farmers who have little access to modern inputs and market outlets, or by commercial companies with predominantly private capital (CCPCs) using farming assets originally distributed to ex-members of cooperatives.

In the pre-reform period, there were also intensive state-owned livestock complexes, mainly in pigs and poultry. They did not have land but were vertically integrated with processing and trading activities. These complexes were burdened with high debts, absorbed a great deal of state transfers, and were difficult to privatize. Many of them were liquidated, and some were privatized as late as in 1999.

Over 60 percent of the country's total land area, that is, about 14.8 million ha, is used for agriculture, mostly arable planted to maize, wheat, and oilseeds. There are also small but important wine and fruit sectors. As in many other Central European countries, livestock numbers and meat production fell drastically during the 1990s, as feed and fertilizer became lacking (the former due to drought as well as loss of farmers' purchasing power) and falling real consumer incomes had a major impact on demand. With poor access to markets and lack of off-farm employment and income, many small family farms have become semisubsistence in nature.

The disintegration between 1990 and 1997 of the formerly highly concentrated upstream and downstream enterprises deeply affected Romanian agriculture. The "agromecs" used to supply farmers with land preparation, crop harvesting, and irrigation services, while the "integrators" supplied farmers with seed and agrichemicals, as well as procuring their crop output. A privatization program has been adopted for both these subsectors in order to stimulate competition in input supply and output handling. However, given the poor state of most of the capital in these subsectors and the slow development of inward investment, there is still much to be done.

External events have also discouraged agricultural development in Romania. The breakup of the Comecon trading bloc and the 1998 economic slump in Russia removed a potential market for some Romanian products. Later, the upstream blocking of the Danube during the allied bombing of Serbia disrupted waterborne transport to the Black Sea and beyond.

In the face of these problems, Romania has remained almost self-sufficient in agricultural and food products (except for sugar). Although imports of products such as higher-quality foods, beverages, and tobacco have risen, less-processed products such as grains, live animals, wine, and oilseeds have been exported. All this suggests an underdeveloped and inefficient food processing and trading sector, which is unable to compete in adding value to domestic farm production. At the same time, it suggests possible opportunities for the future.

Agricultural Policy

Until 1997, when the system was abolished as part of "shock therapy" conditional for a World Bank loan, minimum prices for several farm products "of national importance" (wheat, milk, pork, and poultry) were guaranteed by the state through the "integrators." These were supplemented by input subsidies and by extensive direct payment and credit schemes, mainly to state-controlled enterprises. Indeed, half of the "quasi-fiscal deficits" in the 1996 government budget came from subsidized credit directed by the government to what were essentially non-creditworthy state enterprises in agriculture. Fiscal transfers represented a quarter of the total support for agriculture during 1993–1995 and reached over 40 percent in 1996. Total transfers to agriculture accounted for some 3.9 percent of GDP (Tesliuc 1996). This scale of transfers was clearly unsustainable, undermined monetary discipline and thereby contributed directly to inflation. High nominal interest rates also curtailed both short-term seasonal borrowing and medium- to long-term investment in agriculture, particularly when fuelled by inflationary expectations. At the same time, the overvalued currency hindered exports. In this unfavorable environment, Romania had to proceed with major agricultural reform measures.

Since 1997, agricultural prices have been essentially liberated, subject to occasional trade interventions. Between 1997 and 2000, the government operated a "voucher" system of input subsidies which could be used for various farm inputs such as seed, fuel, and fertilizer. Through the post, small landowners were sent vouchers, worth about $16 (in 1997), according to area up to a limit of 6 ha. These could be used to pay input suppliers, who, in turn, could convert them into cash at authorized banks. Vouchers accounted for the major share of government expenditure on agriculture and became a litmus test for government attitudes toward the agricultural sector according to how eligibility and timing was determined from year to year. In 2001, the system was replaced by a lei 1 million per ha subsidy program (Serbanescu 2002).

The rest of the agricultural budget has been devoted to a number of smaller schemes, such as subsidies for farm credit, for land reclamation in less-favored areas, for irrigation infrastructure, for certified seed, and for a livestock improvement program. Legislation similar to EU regulations is now in place for plant and animal health, food hygiene, soil and water protection, and organic farming. However, the efficiency of the systems required to apply this legislation is sometimes reduced due to lack of equipment, farmer awareness, and administrative capacity. Rural development policy is still in its infancy, as are agri-environmental measures.

The 1991 Land Law, the first step in Romania's land reform program, established the initial parameters for restitution but explicitly prohibited land sales. After long debates and several legal amendments, the so-called Lupu Law no. 1/2000 confirmed the upper restitution limit as 50 ha of farmland (plus up to 10 ha of wooded land), and led to a further and presumably final wave of restitution and/or reimbursements. In the 1991 Land Law, a distinction was made between the so-called state's private estate and state's public estate. The "state's private estate" is subject to privatization and consists of state farms (or Agricultural State Enterprises). Since 1990, they have been mostly transformed into commercial companies (the CCSCs) with large debts, and have undergone much slower privatization. The Agency of State Estates (or State Domain Agency) was set up in 2000 to promote their disposal, but the end of 2001 had privatized only 20 of the 600 CCSCs. The "state's public estate" consists of common pastures, experimental and educational farms, and it is not subject to privatization. Land reform legislation has often not been fully or efficiently implemented in practice, due to weak administrative and judicial systems. This situation has resulted in many informal land-use arrangements (poorly reflected in available statistics) but has discouraged the secure restructuring of farm holdings, investment in farming, and the emergence of an efficiently functioning land market. In 1995, the government amended earlier legislation to make more difficult the privatization of land expropriated by the state in 1945 (e.g., by requiring clarification of legal status), and state-owned land may actually be increasing.

Under the General Agreement on Tariffs and Trade (GATT) and the World Trade Organization (WTO), Romania—uniquely amongst the CEECs—has the status of a developing country. Thus, it is permitted relatively high tariff rates and generous export subsidies (which are expressed in 1986–1989 prices to avoid erosion due to inflation) on its most important agricultural products. However, domestic budget pressures and moves to agricultural trade liberalization have avoided protectionist exploitation of this situation. Since May 1997, the applied tariff rates have generally been a good deal lower than the bound (maximum allowed) levels, averaging 27 percent in 1998, and 22 percent after May 1999, with sensitive commodities not above 40 percent. In 1997, Romania joined the Central European Free Trade Agreement (CEFTA), which stimulated some interregional trade. However, the effect on domestic markets of cheaper wheat and meat products from Hungary led to the temporary suspension of the relevant tariff concessions in 1999–2000, and even to consideration of withdrawal from CEFTA.

Table 1.2 shows the levels of Producer Support Estimates (PSEs) in Romania and other Central European countries. These are regularly calculated by the Organization for Economic Cooperation and Development (OECD) as an indicator of transfers from consumers and taxpayers to farm producers.

Table 1.2 Producer Support Estimates (PSEs) for Romania and Other EU Applicants, 1991–2000 (percentage)

	1991	1997	1998	1999p	2000e	Average 1997–2000
Romania	**15**	**3**	**28**	**18**	**11**	**15**
Bulgaria	–39	–10	2	–6	2	–3
Czech R.	51	4	20	20	18	15
Estonia	59	5	19	5	10	10
Hungary	13	5	19	23	18	16
Latvia	83	4	16	17	18	14
Lithuania	–262	3	13	14	9	10
Poland	–1	14	21	21	20	19
Slovakia	39	8	27	24	22	20
Slovenia	35	37	44	48	43	43

p = provisional
e = estimate

Source: OECD (2001)

The percentage PSE (support compared to output at domestic producer values) averaged 50 percent in the period 1986–1989, but then fell to relatively low levels, and then lower still in 1997 as a consequence of the price and trade liberalization moves of that year. Since then, the percentage PSE has fluctuated with world prices (which are not fully reflected in domestic prices), and a real appreciation in the exchange rate. Though not shown, the Consumer Support Estimates (CSEs) have been generally negative, despite some consumer subsidies, and reflect the implicit tax paid by Romanian consumers for an inefficient food processing and distribution system. Total Support Estimates (TSEs), which add general services and budgetary payments to the PSE measure, remained relatively high among OECD-monitored countries.

EU Accession and Adoption of the CAP

As with the other applicant states of Central Europe, Romania has experienced a sequence of pre-accession developments dating back to an interim EU Agreement in 1993 and the ratification of the main Europe (Association) Agreement in 1995. A revised Accession Partnership was adopted in December 1999, when Romania was invited to open the accession negotiations itself, and since then there has been slow progress in a number of areas. So far, only eleven "chapters" have been provisionally closed in the enlargement negotiations, compared to seventeen for Bulgaria, twenty-three for Poland and twenty-four for Hungary (European Commission[EC], Directorate General [DG] enlargement Web site, 22 April 2002), and the Agriculture chapter has not even been opened.

The impact of the Europe Agreement on EU-Romanian trade has so far been "minor" (OECD 2000), due to a mismatch between the EU tariff quotas and Romania's export potential and to more general exporting problems of quality, credit, etc. On the other hand, Romanian imports of EU agri-food products have increased significantly, though not necessarily as a direct result of the agreement. Reciprocal concessions for agricultural products, including the "double zero" option (the reciprocal opening of duty-free tariff quotas and the abolition of export aids) for wheat, pork, poultry, and cheese entered into force in 2000 and 2001.

In May 2000, Romania presented a revised National Program for the Adoption of the Acquis (NPAA), that is, a strategy for accession including the achievement of priorities in the Accession Partnership. Under the EU's Special Accession Programme for Agriculture and Rural Development (SAPARD)—one of the pre-accession instruments—Romania is due to receive about Euro 150 million per year over the period 2000–2006. Under the National Agricultural Rural Development Plan, the main items in the SAPARD budget are for processing and marketing agricultural and fish products, for investment in rural infrastructure and in agricultural holdings, for farm household diversification, and for forestry.

As regards the instruments of the Common Agricultural Policy (CAP) itself, "there has been little progress as regards the introduction of a legal framework and administrative structures for the establishment of Common Market Organisations" for agricultural products (European Commission 2001, 54). Thus the main commodity-specific devices for administrating and controlling the complex CAP systems of direct payments, marketing quotas, import levies, and export subsidies have yet to be harmonized along EU lines. There have been advances in marketing arrangements for wheat, wine, fruit, and vegetables, but further efforts and changes are needed as regards border inspections and veterinary and phytosanitary systems.

Conclusion

This short introduction to Romanian agriculture and its economic situation over the last decade shows that the country now faces some critical decisions if the declared intention of achieving accession to the EU by the year 2007 is to be realized. Many of the issues touched upon here are discussed in more detail in further chapters of this book. Whatever the reasons for the difficulties of adjusting the economy and the policies of the country to post-1989 conditions may have been in the 1990s, there has now been time to appraise the effects of the slow dismantling of the old state-controlled structures. Among other CEECs, there are also now plenty of examples of more rapid progress toward a market-oriented economy, and much experience and expertise have been built up in the other EU accession negotiations. Although, like all countries, Romania possesses some individual characteristics and problems, there is little reason to use these as an excuse for further delays if EU accession is to remain a medium-term goal.

Like many other countries, Romania's basic policy problem is one of encouraging farm labor to move to off-farm employment. In primary agriculture, returns are poor, and farm level restructuring and improvements are inhibited by small holdings and lack of capital. Apart from general recommendations for better education and transport facilities and efficient public administration, both of which should promote private enterprise in general, the agri-food sector is an obvious target for attention. Reform and progress in this sector would simultaneously provide alternative employment for at least some of the rural workforce, reduce food costs to the poorer sections of the national population and to the state budget, and improve Romania's export performance by providing higher-value processed food and drink products.

In addition to this central recommendation, there are a number of other specific aspects to be considered. Conditions in the domestic Romanian markets for farm products are not only affected by the behavior and performance of downstream sectors but by international trade, which can be quickly destabilizing given the presence of several major agricultural countries nearby. In addition, the climate of Central Europe can markedly alter crop yields (and hence livestock feed costs) from year to year. In such a situation, any set of relatively small countries, with large numbers of food-sensitive consumers, would be likely to find it difficult to operate liberal trade policies. Only a large and economically stable free trade area would reduce the problems of fluctuating trade quantities and prices, and even then some severe farm income crises could be expected from time to time.

In the meantime, Romania, like other Central European countries, must find a way to continue the orientation of its agricultural trade policy instruments toward those of the EU, at the same time as taking advantage of uncertain markets in its natural trading hinterland around the Black Sea and further afield. There may also be problems arising from extra commitments taken on if a move is to be made to WTO-developed country status in 2005, when the current world trade negotiation round is due to be completed.

The difficulties of reforming the land tenure system are also great, given the lack of other attractive long-term assets for the poorer and smaller household. However, without provoking major resistance from those who have only recently recovered effective land rights after decades of state expropriation, it should be possible to develop further the informal farming arrangements already in place. Speedy and rigorous enforcement of land, labor, and product contracts would greatly assist confidence amongst large and small investors, and encourage moves to larger units which should suit Romania's excellent natural conditions for arable agriculture.

It is easy enough to point out the problems of Romanian agriculture and not so difficult to recommend solutions. This does not absolve outside governments, and others, from behaving in ways which encourage solid progress toward economic prosperity and EU accession. With a PSE still around 40 percent, and the huge budget cost of the CAP, the EU is in a weak position to argue for radical agricultural policy reform amongst its neighbors, particularly when its own CAP reform intentions are unclear over the next decade. Undue difficulties should not be placed in the way of accrediting plants, companies, and administrative systems which are genuinely capable of meeting the conditions of EU membership as regards reliable and internationally acceptable food standards. Recent Commission proposals (Draft Common Positions, Autumn 2001; see Agra Food *East Europe*, September 2001, 1) to permit derogations from veterinary and health criteria to certain farms and food-processing units in applicant countries should lessen the chances that accession is pushed further and further into the future on this account.

Another consideration facing Romania is that, on present trends, several other CEECs will join the EU before its own accession is possible. The current members may need to exercise themselves in ensuring that the newest entrants do not exploit their voting power to hinder progress by those behind them in the accession queue. Simultaneously, Romania must continue to press its claim for full consideration as an eventual member, taking advantage of available EU funding under the PHARE, SAPARD, and ISPA (Instrument for Structural Policies for Pre-Accession) programs, as well as exploiting available tariff quotas into the EU and other markets.

The EU Enlargement Commissioner, Günter Verheugen, has said recently that, despite serious difficulties, Romania's membership is "no longer a question of 'if,' but of 'when.'" The date and success of that membership is heavily dependent on the performance of Romanian agriculture over the next few years. In turn, that performance depends both on the individual efforts of Romanian farmers and agribusiness personnel, and on the ability of the Romanian government to provide both stability and a constant set of incentives toward economic efficiency. This book contains a number of studies dedicated toward several of these challenges.

References

Agra Food *East Europe*, No. 228. September 2001, 1

European Bank for Reconstruction and Development (EBRD). 1998, 1999, 2000 *Transition Report.* London: European Bank for Reconstruction and Development.

European Commission. 2001. *2000 Regular Report on Romania's Progress to Accession.* Brussels: Enlargement Directorate-General.

Organization for Economic Cooperation and Development (OECD). 2000. *Review of Agricultural Policies: Romania.* Paris: Organization for Economic Cooperation and Development.

OECD. 2001. *Agricultural Policies in Emerging and Transition Economies.* Paris: Organization for Economic Co-operation and Development.

Serbanescu, C. 2002. *Review of Agricultural Policy, Markets and Trade Developments in Romania in 2001.* Global forum on Agriculture: Workshop on Agricultural Policies and Support in Non-member Transition Countries, Paris.

Tesliuc, E. 1996. *Regimul comercial cu produse agricole in Romania.* Bucharest, Romania: World Bank Resident Mission. Mimeo-graphed.

Zecchini, S. 1997. Transition approaches in retrospect. In *Lessons from the Economic Transition: Central and Eastern Europe in the 1990s*, ed. S. Zecchini, 1–34. Dordrecht/Boston/London: Kluwer Academic Publishers.

Chapter 2

Agricultural and Trade Policies

Sophia Davidova and John Lingard

Since Romania started to move on the path of economic restructuring in 1990, its reforms have been subject to much criticism by internal as well as external experts and by international financial institutions. The main points of criticism have been the slow speed and inconsistency of the reform process: the approach has been described as "stop-go" (Balcerowicz and Gelb 1995, 21–44). Despite the general disappointment, a large number of changes have been introduced in agricultural and trade policies.

Chapter 1 outlined some of the main agricultural policy developments in Romania during transition. This chapter tries, first, to go deeper into price and trade policy reform in Romania and, second, to analyze some of the policy effects with a particular emphasis on the switch to more liberal policies in 1997.

Domestic Agricultural Support Policies before 1997

The government objectives for the agriculture and food sector in Romania before 1997 were, broadly speaking, to keep consumer prices low while attempting at the same time to provide incentives to producers. Romanian policies were highly distortive and relied heavily on price intervention. Maintaining employment within agriculture and the parastatals in the food chain was also an implicit and much-practiced objective, along with ensuring food security and regional development goals.

The main policy instruments used in Romania up to 1997 were price interventions for wheat, pork, poultry, and milk. Prices were fixed at the farm-gate, and margin control was extensively exercised throughout the agri-food chain for these four strategic commodities, which accounted for 23 percent of total consumer expenditure. Producers selling to state-mandated processors at the fixed farm-gate price were eligible to receive "premia" amounting to one-third of the fixed price. The premia played the role of deficiency payments and were not included in the costs of downstream sector. A large share of the marketed output of these four commodities was sold through the state-administered marketing chain. Some sales at market prices occurred, mainly for wheat and milk. In general, administered prices allowed little scope for seasonal and regional arbitrage; there was no private entry into this business, and price competition had been completely eliminated.

The intended effect on consumer prices was achieved for bread, milk, and dairy products, but not for meats. The average food price index in 1995, based on October 1990, was 10,452. The indices for bread and for milk and dairy were below the average at 7,688 and 9,937, respectively, while the meat index was 12,407 (Davidova and Ivanova 1997).

Price intervention was supplemented by input subsidies and direct budgetary support to producers and intermediaries. Direct credit schemes were used extensively, mainly to the benefit of CCSCs, state-owned intermediaries, and processors. Total budgetary support for agriculture in 1996 was lei 2,182 billion; price premia accounted for most of this, lei 1,217 billion, and interest rate subsidies were lei 341 billion. Other subsidies funded half of the price of certified seeds and also the free distribution of about 40–60 kg of fertilizer/ha to grain producers, who signed contracts with state-mandated agents to supply at least 40 percent of their wheat crop at the administered price. Running costs of the public irrigation system were also funded from the state budget. Indirect tax support was offered to agriculture in the form of profit tax relief.

In order to estimate the overall impact of the contradictory policy instruments on producers, it is necessary to calculate some of the measures of support and protection. These measures are well known, along with their strengths and weaknesses. Here the Nominal Protection Coefficient (NPC) is applied. The NPC compares the opportunity costs of domestic production, using the border price as a proxy, to the price domestic producers face (Tsakok 1990). In practice, it is a comparison between the domestic price of agricultural output and the border price. In order to compare both at the same level of the food chain, either the border prices are adjusted back to the farm level or the farm level price is adjusted to the border. If the NPC is smaller than one, then domestic producers are being taxed, as they are operating below their opportunity costs. If the NPC is larger than one, then producers are being subsidized. An NPC equal to one indicates a lack of distortions and operation at competitive prices. However, it should be noted that even when this simplest measure of protection is calculated, there are many problems. Depending on decisions about the time period, level of the food chain, border price, and foreign exchange rate, the results can be quite different. Sometimes it is even difficult to come to a firm conclusion as to whether farmers have been taxed or subsidized. The Romanian example is instructive about the difficulties in calculating protection measures and emphasizes the caution with which such measures should be interpreted.

There are three different sets of measurement of nominal protection coefficients in Romania for the period preceding the liberalization of price and trade policies in 1997. Two of them were made at more or less the same time by different teams working for the World Bank. One of these teams was the Agricultural Structural Adjustment Loan (ASAL) team (Deaconescu, Tesliuc, and Gordon 1996), and the other was the special mission for estimation of effective protection (Metzel and Salinger 1996). The third set of protection indicators were calculated by Davidova and Deaconescu (see Davidova 1997, 131–159). Only calculations using the nominal exchange rate and the official minimum guaranteed prices, including premia, are shown and discussed in this chapter (table 2.1). The three sets presented use the same domestic prices and exchange rate. The main differences between the sets lie in the way in which some calculations were made, in the period covered, and how this affected the choice of border price.

In the calculations, the ASAL, and Davidova and Deaconescu subtracted domestic marketing margins from the border price to derive an equivalent price at a farm level, while Metzel and Salinger built up domestic costs from the farm to the border. As mentioned above, both methods are valid. There were also some small differences in the estimation of marketing costs.

On the period covered, the World Bank approach was to calculate monthly NPC, while Davidova and Deaconescu attempted to present annual averages. Each approach has its strengths and weaknesses. While the monthly NPC could be calculated with more accurate data collected from the main market agents, interpretation of a single-point NPC is risky (Metzel and Salinger 1996). The use of different periods resulted in the application of slightly different exchange rates and marketing costs, but most importantly in differences in the border price. To illustrate this, the wheat price for 1996 is used as an example. ASAL took a free on board (f.o.b.) Constanza price for July just before the new harvest, when the price was at a high level ($204 per tonne), and compared it to the domestic administratively fixed price, which was not subject to monthly adjustments. Metzel and Salinger used the f.o.b. U.S. Gulf September price for soft red winter wheat at $170 and justified this choice by arguing that this is the time of year when Romanian traders have to make export decisions. Davidova and Deaconescu used 1996 export parity unit values, $181 per tonne. Understandably, different approaches yielded different results.

Table 2.1 Romania NPCs for Produce Subject to Intervention before the 1997 Policy Reform

Commodity	NPC						
	Davidova and Deaconescu (Davidova 1997)			ASAL (Deaconescu, Tesliuc, and Gordon 1996)			Metzel and Salinger (1996)
	1994	**1995**	**1996**	**July '94**	**July '95**	**July '96**	**Sept. '96**
Soft wheat	1.40	0.90	1.10	1.34	0.81	0.91	1.20
Pork	1.99	1.89	1.78	1.92	2.00	1.36	1.26
Poultry	1.77	1.28	1.49	1.69	1.34	1.15	1.10
Milk	0.81	0.64	0.74	0.67	0.54	0.64	0.87

The most important difference in the results occurred for wheat farmers in 1996. For 1996, the ASAL approach yielded an NPC of 0.9 for soft wheat. Metzel and Salinger's calculations resulted in an NPC of 1.2, while Davidova and Deaconescu estimated an NPC of 1.1. Therefore, two of the calculations show that at the nominal exchange rate, wheat farmers were slightly supported, while the

ASAL approach resulted in a conclusion that they were slightly taxed. All three sets confirm that the milk producers had been taxed.

These calculations show that wheat, pork, and poultry farmers need to increase their efficiency in order to survive under liberal policies. However, milk producers could only gain from more market-oriented and less consumer-targeted policies. Therefore, the major policy reform that followed in 1997 implied different effects for different groups of farmers, depending on their output mix.

The 1997 Policy Reform and the Subsequent Support Measures

Overall, the price and trade policies in Romania during the first years of transition were highly distortive. They were counterproductive to the reform process, as they sheltered the large-scale loss-making CCSCs and state integrators. As pointed out by OECD (2000, 98), "the policies responded poorly to a major shift in the production structure that increased the importance of small producers." The change in government in late 1996 coupled with external pressure from international financial institutions made the policy liberalization feasible.

Agricultural Policies

Continuing with the past policies was unsustainable from the point of view of the budgetary burden, outlined in chapter 1, and their role as an impediment to restructuring of agricultural sector, and the 1997 reform program explicitly recognized this. The national economic program was set out in February 1997 after negotiations with the International Monetary Fund (IMF) and the World Bank. Price and margin controls were removed from all agricultural products, and key reform measures were announced (box 2.1). The government's role was thus no longer to be seen as acting as a substitute for an imperfect market system, but as a complement to stimulate and facilitate market development (Lingard 1998).

Box 2.1 Measures Included in the National Economic Program in Romania, February 1997

- To abolish administrative price setting through the agri-food marketing chain and to abandon premia subsidies.
- To liberalize the trade regime.
- To compensate the low-income food consumers for the effects of food price liberalization.
- To reform the cereal sector and to privatize the grain companies within the National Agency for Agricultural Products (ANPA).
- To ensure food security and to create a foreign exchange fund to be used solely for emergency food imports.
- To accelerate the privatization and liquidation of nonviable enterprises, including CCSCs, specialized state-owned seed farms, pig and poultry complexes.
- To phase out the state-directed subsidized credit and to encourage private banks to reach out to agriculture and rural enterprises.
- To develop a rural financial system.
- To amend land legislation and to ensure free land sales by individuals and the development of a land-leasing market.
- To reduce the direct role of the MoAF in agricultural production, processing, and marketing, and to focus its role on the provision of public goods to agriculture.

The ASAL was signed with the World Bank in 1997. The aim of the ASAL has been "to introduce efficient market incentives, increase private sector entry and competition, and reduce direct Government intervention in agriculture" (Report and Recommendation 1997). The amount was negotiated at $350 million, which had to be disbursed in two tranches.

The structural elements of this program concerning privatization and the land market were delayed, with the result that in 1998 the World Bank was reluctant to disburse the second ASAL tranche. However, price and trade liberalization was implemented in 1997. Agricultural producer and consumer prices were effectively liberalized in mid-February, and the law providing for price intervention was abolished. This was coupled with the liberalization of nonfood prices, where the greatest implications were from the liberalization of fuel and energy prices. Price liberalization included the removal of the minimum guaranteed prices and premia, as well as of all controls on profit margins.

The immediate effect was a short-term increase in consumer prices. However, the first price shock was not followed by subsequent increases. Within four to five months, prices started to decrease with the drop in demand. For example, the price of standard white bread (which is the most commonly consumed type of bread in Romania) initially increased by about 150 percent in comparison with the administratively controlled preliberalization price. It stayed more or less at this new level for three months and then began to decline after the new harvest (declining by 15 percent at the beginning of August 1997). Dairy and meat prices followed a similar pattern. A six-month scheme for compensation of the most vulnerable consumers was introduced, amounting to $119 million.

Farmers' prices for produce supported before the reforms fell after the policy liberalization. The biggest problem that arose related to the wheat price. At the end of July 1997, the price of soft wheat was between $100 and 107 per tonne. This was the price level which allowed exports, since the July-August border parity prices (f.o.b. Constanza) were about $120–125 and the internal transport and marketing costs to the border were $20–$25. Farmers complained that the new price did not cover their production costs. In order to smooth the adjustment process, the government continued some support schemes, for example, seeds and storage subsidies. However, gradually, these were either downscaled or phased out. The main schemes that remained were credit subsidies and the newly introduced input vouchers.

Some transparent directed credit funds were included in the state budget for spring planting in 1997 and purchase of the summer harvest (lei 550 and 500 billion, respectively), and were rolled over for the autumn planting. The state budget lent to commercial banks that would bear the full risk of lending. A subsidy of 15 percent of the lent and recovered principal was paid by the government to encourage commercial banks and other financial institutions to participate.

The traditional way of channeling input subsidies through the state-mandated intermediaries was abolished and instead a voucher scheme for inputs was introduced. This scheme was designed by the World Bank as a minimally distortive instrument. Under the scheme, all landowners received vouchers directly to enable them to purchase inputs. Vouchers have been distributed to landowners on the basis of their landholding with a minimum threshold of 0.5 ha. CCSCs were not eligible for support. Members of ASs and FAs who owned land also received vouchers. Therefore, the first effect of the scheme was to support private farming and to use policy discretion to pressure the CCSCs to restructure. The second main effect was to switch input marketing from former state integrators to private channels of wholesalers and retailers and to provide farmers with a wide range of inputs and technology supplied by competitive, private markets. Vouchers could be used to purchase seed and other planting material, fertilizers, and pesticides, and to pay for machinery services. Inputs could be bought from any supplier and will, thus, in theory, "kick-start" the development of private input channels and increase market competition.

The voucher scheme was designed to be temporary for two to three years only and was to be phased out once distribution networks for inputs had become established and the emerging input marketing sector had become strong and flexible enough to support itself. However, the scheme continued with some modifications in 1999 and 2000. One of the main modifications was to include livestock producers and producers of sugar beet, grapes, vegetables, and tree fruit in the group of eligible farmers. The number of vouchers per hectare was differentiated according to the type of production. The vouchers distributed in 2000 had a nominal value of lei 100,000 each (about $5) (Serbanescu 2001). The highest number of vouchers per hectare, twelve, was allocated to vineyards and orchards, and the lowest, five, to cereals and oilseeds. Animal producers benefited from four vouchers per head of dairy cow or dairy water buffalo. The value of vouchers distributed in 2000 was lei 4,853.4 billion, or about $255 million, accounting for more than half of the MoAF budget for 2000. Interviews with farmers during summer 2000 showed that, on average, vouchers covered 29 percent of the costs of variable inputs used. Analysis of the effects of the voucher scheme on different management types of farms is reported in chapter 6.

The voucher scheme was practiced for four years and then discontinued in 2001. The idea was to replace it with a system closer to the CAP; namely, direct payments set at lei one million ($35) per hectare. All crop, horticulture, and viticulture products were eligible for support conditional on a minimum size area (1 ha, except for sugar beet and vegetables, 0.5 ha) and appropriate technology, including the use of certified seeds, application of fertilizers, etc. (Serbanescu 2002). The total amount spent on this support in 2001 was $193 million.

Although Romania greatly liberalized its agricultural policies in 1997, in the following years some of the previous distortive support instruments have been reintroduced on an ad hoc basis, following changes in the government and the management of the MoAF. These support instruments can be classified into three groups. The first includes price premia for sales of wheat and milk to licensed economic operators. The second group is comprised of credit and other input subsidies. The credit subsidies were reinstated through an earmarked fund created in 1998 and incorporated in the budget of MoAF. Subsidies for short-term credit applied to credit with a one-year maturity. In the case of due repayment, farmers received a 70 percent interest rate subsidy. A similar scheme was introduced for midterm credit (up to five years). In addition to the interest rate subsidy, subsidies included a one to three years' period of grace. The estimate of the MoAF budget allocated for interest subsidies on short- and midterm credit was lei 49.5 billion (or about $2.3 million) by mid-2000. The third group of intervention measures includes investment support through the fund "Development of Romanian Agriculture." This fund was supposed to be based on allocations from the Agency of State Estates set up in 2000. The agency was responsible for finalizing land reform and for providing concessions on state land and assets. The investment support was targeted at the purchase of machinery, breeding animals, and irrigation equipment. Some of the programs are strongly reminiscent of the time of central planning. A list of selected policy measures is presented in table 2.2.

In conclusion, after the liberalization of agricultural policy in 1997, some distortive schemes were continued. However, instead of being treated as transitory with the objective of facilitating producers' adjustments and giving political credibility to the policy reform, they have become a permanent feature of Romanian policy, and new distortive measures have subsequently been introduced. This has continued the existence of soft budget constraints and delayed the expected positive effects of the agricultural policy reform for economic efficiency.

Table 2.2 Agricultural Policy Instruments Applied in Romania, 2001

Content	Eligibility
Market Price Support	
Price premium for wheat, $20/tonne	Producers selling bread wheat for domestic bread consumption to mills, bakeries, and silos
Price premium for milk, $17/tonne	Producers selling milk to licensed processors
Direct payments	
Direct payment for land cultivation, $35/ha	Producers, individuals, or legal entities with private or mixed capital, research and education farms in crop, horticulture, and viticulture; conditional on minimum area and technological requirements
Credit subsidies	
Interest subsidies for credit for working capital	All types of producers with private or mixed capital; list of crops and livestock eligible for support
Interest subsidies for purchase and storage of wheat	Bakeries, feed mills, livestock operators purchasing bread wheat or feed barley with quantity limits
Subsidized machinery purchases	
Subsidy equivalent to 55 percent of the machinery price	All producers except CCSCs and providers of agricultural services for purchases of domestically manufactured machinery
Input subsidies	
Diesel subsidies for agricultural campaigns, price of diesel without tax	Private agricultural producers cultivating area over 0.5 ha
Certified seeds subsidies, paid to certified seed producers (between 40 and 50 percent of the price)	Agricultural producers purchasing certified seeds

Source: Based on Serbanescu 2002

Trade Policies

Before 1997, trade policy in Romania was characterized by a high rate of import protection and the simultaneous use of export impediments (such as bans, quotas, and registration requirements). The import policy was targeted at supporting farmers, while the export restrictions had the opposite objective, namely, taxing producers and supporting consumers. Within this framework there were also the so-called administrative waivers. These were low or zero rate quotas for imports allocated to specified users, for example, imports of barley for the state pig and poultry units. In general, the trade policy pursued in Romania was adverse to the creation of an internationally competitive agriculture. The tariff protection was not challenged internationally because of the high WTO bindings (see chapter 12).

The reform of trade policy included the liberalization of agricultural exports and imports by eliminating existing impediments and decreasing the import tariffs. Export impediments were removed, as were the administrative waivers for imports. The ASAL target for import liberalization was to set the maximum tariff at 60 percent and to achieve a production-weighted average of agri-food tariff protection of 27 percent in 1997. This rate was to be decreased to 22 percent in 1998. The World Bank excluded from the liberalization requirements chapters 22 (beverages) and 24 (tobacco and tobacco products) of the Harmonized Customs System (HS). The changes in tariff protection due to policy reforms in 1997 are shown in figure 2.1, which presents production weighted averages for raw agricultural products, processed food products, and for the agri-food sector overall (data include beverages and tobacco).

In 1996, the production-weighted average was 102 percent. After the 1997 liberalization, it fell to 33 percent. One of the main changes in the relative tariffs in 1997 was that, while in 1996 the border protection of agriculture was higher than that for food processing, after the liberalization, it was the opposite.

Figure 2.1 Romania, Production Weighted Average Import Tariffs, Ad Valorem

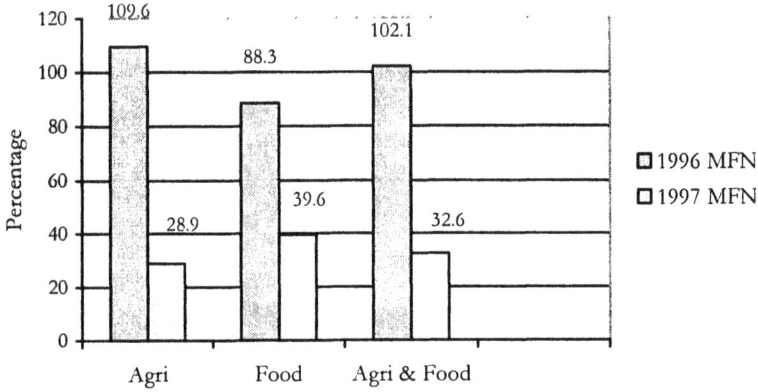

Source: Own calculations based on Tesliuc (1996); Romanian Customs Tariffs (1997)

Overall, with the exception of the tobacco and beverages, the maximum applied Most Favored Nation (MFN)[1] tariff was set at 60 percent. Romania thus met the ASAL conditions. By forcing the maximum tariff to 60 percent, the ASAL brought about another positive development: a decrease in tariff dispersion. The variation coefficient for the tariffs covering the whole agri-food sector was 991.4, while for the chapters subject to ASAL conditionality (as mentioned, the conditionality does not apply to beverages and tobacco) it was 277.2. For the two product groups for which Romania has not been asked to liberalize as a condition for the ASAL, however, the maximum tariffs were set much higher than 60 percent: 248 percent for beverages and 98 percent for tobacco.

The livestock sector received more protection than the crop sector. In this way, Romania tried to protect its politically and socially sensitive and internationally uncompetitive production of pork and poultry. The highest protection within crop products was granted to vegetables, for which Romania is a traditional producer.

The trade policy reform in mid-1997 coincided with the accession of Romania to CEFTA. The lists of bilaterally granted concessions for "sensitive" agricultural products are relatively short. The bilateral concessions are spread over several product groups but, with the exception of fresh vegetables, most of the concessions are for processed products (Agreement on Accession 1997). The multilaterally agreed trade liberalization with CEFTA is substantial. The production-weighted average in 1997 was 21.2 percent compared to 32.6 percent for MFN. The relative tariffs are similar to MFN, and protection of the food industries is higher than that for agriculture.

In summary, in 1997 Romania made a decisive step in trade liberalization, particularly in decreasing tariff protection. This has been a step toward increasing import competition and inducing pressure on domestic producers to increase efficiency. However, there have been strong political pressures to again raise import tariffs for some products and to allow zero tariff imports for other products on an ad hoc basis. To the extent that WTO bindings are so high that they allow increases in applied tariffs, the internal political economy is the main factor in protection decisions.

The main political sensitivity in Romania has been the imports from Hungary, due to its geographical proximity and similarity in the structure of output. In 1999, Romania suspended CEFTA concessions granted to Hungary for live pigs, pork, and meat products, and reintroduced import tariffs of 15–20 percent for live animals and 45 percent for pork and chicken meat. The validity of these measures was prolonged for 2000 (Serbanescu 2001). Romania also reintroduced its long-lasting policy for temporary import tax exemptions. For example, in 2000 such exemptions were in place for a quota of 500,000 tonnes of maize, for barley, molasses, and other products. This reintroduces the instability in import rates for some of the main products that potentially could have negative effects on the policy environment for the development of agriculture.

The overall effect of agricultural and trade policy reform and the subsequent reintroduction of distortive intervention measures can be exemplified by the PSE calculated by the OECD. As discussed in chapter 1, PSE is an indicator of the annual monetary value of gross transfers from consumers and taxpayers to support agricultural producers, measured at farm-gate level. PSE can also be expressed as a percentage. The percentage PSE is the total value of transfers expressed as a percentage of the value of production. The higher the percentage, the higher the support provided to agricultural producers through prices or through budgetary funds. From table 1.2, it can be seen that in 1997 in Romania support dropped after the price and trade liberalization to a record low level of 3 percent. But this was short-lived. As discussed, the government reintroduced various support measures, and in 1998 the percentage PSE increased to 28 percent (see table 1.2). However, it should be underlined that the fluctuation in world prices and changes to the exchange rate are factors that influence the PSE, and thus it does not only reflect the changes in domestic agricultural policy.

A key principle of agricultural policy reforms is to let market signals influence the orientation of agricultural policies. This is important for all countries in the world, but even more for the countries in transition, including Romania, as the core of the transition to the market is to move from a system responding to planned targets and shortages (Kornai 1980) to a system responsive to market signals. The responsiveness to markets (prices) is crucial for inducing efficiency and competitiveness in the agri-food sectors in the countries in transition. In order to allow market signals to influence the orientation of agricultural production, it is necessary to reduce support to the sector and to deliver support through less distortive measures. Romania has not sustained this policy and quickly went back to its typical ad hoc policy, often employing contradictory instruments lacking internal consistency.

Conclusion

Romania started an accelerated move toward more liberal agricultural and trade policies in 1997. However, the lack of structural reforms, or at least reforms with adequate speed, dragged Romania into a deeper recession in 1998. The wholly unfavorable macroeconomic environment has left little scope for the fine-tuning expected from agricultural policies.

Agricultural policy and trade reforms in Romania in 1997 show the importance of external pressure for radical reforms in countries in which the internal political economy does not really prompt such reforms. However, policy liberalization was short-lived. This was coupled with a delay in the structural part of the reform program, which has not allowed the agricultural sector to benefit from the less-distortive policies introduced in 1997.

Trade liberalization has survived despite the strong vested interest pressure for increases in import tariffs and for the introduction of some temporary import tax exemptions. Some have been introduced, but on a smaller scale compared to the pre-1997 period. The external constraints put on by ASAL conditionalities have been decisive in keeping Romania on the road of trade liberalization.

Acknowledgments

The research assistance of Virginia Campeanu, Institute of World Economy, is gratefully acknowledged.

Notes

1. These are nondiscriminatory tariffs applied to countries with which there are no special preferential trade agreements.

References

Agreement on Accession of Romania to the Central European Free Trade Agreement. 1997.

Balcerowicz, L. and Gelb, A. 1995. Macropolicies in transition to a market economy: A three-year perspective. In *Proceedings of the World Bank Annual Conference on Development Economics 1994*, 21–44. Washington, D.C.: World Bank.

Davidova, S. 1997. Bulgaria and Romania en route to CEFTA and their Agricultural Policies. In *Regional Trade Arrangements and Agricultural Policy in Central and Eastern Europe*, 131–159. PHARE Multi-country Trade Development Programme and Wye College. Wye, Ashford: Wye College Press.

Davidova, S. and Ivanova, N. 1997. Study on Bulgaria and Romania. In The CAP and enlargement: agrifood price developments in five associated countries. *European Economy*. 2:105–170

Deaconescu, C., Tesliuc, E., and Gordon, H. 1996. *Producer Price Intervention and Incentives in Romanian Agriculture* (draft for discussion). World Bank Resident Mission, Bucharest, Romania. Mimeo-graphed.

Kornai, J. 1980. *Economics of Shortage*. Amsterdam, New York, Oxford: North Holland.

Lingard, J. 1998. *Instruments for Agricultural Policy in Romania.* PHARE project EU Alignment and Policy Reform Report. Mimeo-graphed.

Metzel, J. and Salinger, L. 1996. *Agricultural Effective Protection Analysis.* World Bank Resident Mission, Bucharest, Romania. Mimeo-graphed.

Organization for Economic Cooperation and Development. 2000. *Review of Agricultural Policies: Romania.* Paris: Organization for Economic Cooperation and Development.

Report and Recommendation of the President of the International Bank for Reconstruction and Development to the Executive Directors on a Proposed Agriculture Sector Adjustment Loan to Romania. 1997. Mimeo-graphed.

Romanian Customs Tariffs 1997 and 1998.

Serbanescu, C. 2001. *Review of Agricultural Policy, Markets and Trade Developments in Romania in 2000.* Paper presented at OECD Forum on Agricultural Policies in Non-Member Countries.

Serbanescu, C. 2002. *Review of Agricultural Policy, Markets and Trade Developments in Romania in 2001.* Global forum on agriculture: workshop on agricultural policies and support in non-member transition countries. Paris.

Tesliuc, E. 1996. *Regimul Comercial cu Produse Agricole in Romania.* World Bank Resident Mission, Bucharest, Romania. Mimeo-graphed.

Tsakok, I. 1990. Agricultural Price Policy: A Practitioner's Guide to Partial-Equilibrium Analysis. Ithaca, N.Y.: Cornell University Press.

Chapter 3

The Growth of Individual Farming in Romania

Marian Rizov and Johan F. M. Swinnen

The land reform and farm restructuring that has taken place in Romania since the start of transition was introduced in chapter 1 and forms the background to this chapter. Individual farms in Romania before the 1989 reforms used around 12 percent of total agricultural land (TAL). Since then, the privatization of agricultural production assets and the restructuring of the large-scale state and collective farms have resulted in strong growth of individual farming. By 1997, individual farms used almost 60 percent of TAL.

In comparison with other CEECs, the share of TAL farmed by individual farms in Romania is quite high. For example, the extent of the shift to individual farming is considerably less in the Czech Republic and Slovakia, where individual farms cultivate, respectively, 25 percent and 7 percent of TAL. In Bulgaria and Hungary, individual agriculture accounts for about 50 percent of TAL. Furthermore, considerable differences can be observed across regions in Romania.

Economists, using both a macroeconomic approach and micro-economic survey analysis, have studied the reasons for these differences in the growth of individual farming since 1989. The incentive to start up an individual farm is generally determined by the trade-off between the advantages and disadvantages characterizing individual farming and those of pooling assets into a cooperative farm (Mathijs and Swinnen 1998).

The advantages of individual farming include the lower transaction costs associated with the monitoring of labor and reduced inefficiencies due to the right of codetermination (Lin 1988; Schmitt 1991). Potential disadvantages include the loss of economies of scale in risk management, the provision of information and credit, input purchasing, and marketing and production (Putterman 1985; Carter 1987; Pryor 1992; Deininger 1995).

These costs and benefits of individual farming are also affected by the characteristics of members of the household, such as their age, education, and experience, which are reflected in their managerial capability, and hence these household characteristics are also factors determining the outcomes of the restructuring process (Rizov et al. 2001).

This chapter presents an overview of the development of individual farming in Romania and discusses determinants of the differences observed across regions and households. The discussion draws heavily on a recent econometric analysis by Rizov et al. (2001) who used a 1996 country-representative survey of Romanian rural households.

Farm Restructuring in Romania

Despite vigorous attempts at industrialization during the period of central planning, Romania retained a strong rural culture, with the agricultural sector central to its economy. Agricultural employment accounted for around 30 percent of total employment in the 1980s. In addition, a large number of nonfarmworkers farmed part-time.

Until 1989, agricultural production was organized in three kinds of enterprises: state farms, cooperatives, and small private individual farms (table 3.1). Cooperative farms were created after World War II through the forced collectivization of the assets and land of individual farmers.

By 1989, there were about 3,800 cooperative farms, with an average size of around 2,400 hectares, using 59 percent of agricultural land. State farms were large-scale, capital-intensive enterprises organized on land that had been confiscated from the royal family and large landowners after World War II. Additionally, cooperative farms close to urban centers that suffered from insufficient labor resources were later converted into state farms. Furthermore, new state farms were created on land gained by large-scale drainage and land improvement systems along the Danube.

Table 3.1 Structure of the Agricultural Sector in Romania, 1989 and 1997

Farm Type	Share of TAL[1] (%)	Number of Farms	Average Size (ha)
1989			
State farms	29.0	411	5,001
Cooperatives	59.0	3,776	2,374
Individual farms	12.0	1,000,000[2]	0.5
Total	100.0		
1997			
Commercial companies[3]	12.0	490	3,657
Agricultural societies (ASs)	11.8	3,875	451
Farmers' associations (FAs)	8.4	12,089	103
Individual farms	58.7	3,715,396	2.3
Other institutions[4]	9.1	n.a.	n.a.
Total	**100.0**		

Note: n.a. = not applicable
1 Total Agricultural Land
2 Approximately
3 Predominantly CCSCs
4 Land belonging to agricultural research networks and local councils is included here.
Source: European Commission, DGAgri (1998)

In the pre-reform period, the state controlled most of the agricultural land by means of subjecting most land to central planning. The share of individual farming averaged around 10 percent in 1985,[1] although it varied considerably between counties, ranging from less than 0.1 percent in Teleorman county to 40.7 percent in Hunedoara county (table 3.2). Individual farms were only allowed to remain in hilly and mountainous regions and in remote areas where access to market outlets was often unfavorable. Typically, individual farms were more labor-intensive than cooperative and state farms.

Since the breakdown of the socialist system in 1989, important changes have taken place in landownership and the structure of farms. As outlined in chapter 1, cooperative farms have been transformed into either "agricultural societies" (ASs), employing both member and nonmember workers and applying a cooperative type of management, smaller "farmers' associations" (FAs), or independent farm businesses in the hands of individuals or new companies (the CCPCs).

Table 3.2 Share of Total Agricultural Land (TAL) used in Individual Farms, by county, Romania[1], 1985 and 1993 (percentage)

County	1985	1993
Hunedoara	40.7	97.0
Maremures	31.2	99.1
Caras Severin	28.8	99.0
Suceava	26.8	98.4
Arges	17.0	83.0
Vrancea	15.2	74.6
Neamt	13.3	73.6
Harghita	12.3	94.3
Bihor	11.3	81.3
Covasna	7.8	94.7
Satu Mare	6.5	80.0
Mures	5.8	79.5
Buzau	5.5	49.0
Timis	2.9	72.7
Constanta	0.7	30.4
Tulcea	0.3	26.9
Botosani	0.3	25.9
Galati	0.2	13.1
Ialomiţa	0.1	1.0
Calarisi	0.1	12.0
Teleorman	0.03	5.0

[1] The counties are listed in descending order according to 1985 land share.
Source: National Commission for Statistics (1986, 1994)

Land used by the former cooperatives was partly distributed among cooperative members and partly restituted to former owners, mostly rural households. However, the individualized property rights on land are incomplete. While land in most categories can be leased or sold, co-owners, neighbors, and local authorities have the first right to purchase. Transactions involving a legal person require the approval of the MoAF and the Ministry of Environment. Additionally, changes in land use require the prior approval of the authorities and, if agricultural land is left uncultivated for two years, then the authorities have the right to expropriate it. In 1991, when many of the cooperatives were spontaneously broken up by members, most of the cooperative farm assets were subject to "spontaneous privatization" (Jackson 1997).

According to European Commission data, in 1992 around 1.5 million households out of the 5.3 million that benefited from land restitution or distribution had decided to allocate at least part of their resources to one of the ASs or FAs. At this time there were approximately 4,000 ASs and 11,500 FAs. ASs and FAs not only provide employment for rural workers and cultivate the land but also provide benefits to members in the form of joint transactions with up- and downstream companies (for example, banks and input suppliers).

By 1997, more than 3.5 million individual farms with an average size of 2.3 hectares dominated Romanian agriculture, occupying approximately 59 percent of TAL (see table 3.1). The number of ASs was relatively stable at nearly 4,000, and these had an average size of 451 hectares and occupied a little under 12 percent of TAL. FAs had increased in number to just over 12,000 in 1997. They were smaller (on average 103 hectares) than ASs and occupied just over 8 percent of TAL. The 490 commercial companies that had largely inherited the state farms had an average size of about 3,700 hectares and occupied 12 percent of TAL.

Hence, individual farming has grown rapidly in Romania during transition. Yet there are important regional differences. For example, data from the Romania Statistical Institute show that by 1993 in 25 percent of the counties more than 90 percent of TAL was used by individual farms (see table 3.2). However, in 20 percent of the counties less than 20 percent of TAL was used by individual farms.

Data from 1996 (MoAF/World Bank/EC 1997) and 1998 (Policy Research Group 1999) countrywide statistically representative farm household surveys largely confirm these developments. Furthermore, they yield additional insights. The data show that by 1996 the majority of rural households were engaged in some form of individual farming (table 3.3). Following a classification based on both land and labor allocation, five groups of farm households were distinguished: full-time individual farmers, part-time individual farmers, hybrid farmers, households farming only in association, and absentee landowners.

Table 3.3 Distribution of Rural Households among Five Production Modes, Romania, 1996 and 1998 (percentage)

Household Organization Mode	1996	1998	Average over 1996–1998
Full-time individual farmers	59.8	58.7	59.3
Part-time individual farmers	16.2	18.1	17.2
Hybrid farmers	15.9	20.4	18.1
Association farmers	4.6	0.4	2.5
Absentee landowners	3.5	2.4	2.9
Total	100.0	100.0	100.0

Source: Rizov and Swinnen (2002), based on survey results of MoAF, World Bank, and the EC (1997) and PRG (1999)

Those households that had allocated all their land in an individual farm and where both the household head and his or her spouse were involved full-time in farming were defined as "full-time individual farmers." They represented 59 percent of the total number of farm households over the period 1996–1998. "Part-time farming households," which used all their land themselves, but where both the household head and his or her spouse were engaged part-time in farming, represented 17 percent.

"Hybrid farmers" were defined as those farm households that were employed full-time in agriculture, but that had allocated a part of their assets to a cooperative farm and retained the remainder in their own individual farm. They represented on average 18 percent of the total number of rural households over the period 1996–1998.

The group of farm households that had allocated all their land in a cooperative farm and did not engage in any individual farming, but were employed full-time in agriculture, consisted of around 5 percent in 1996. Over the two-year period between the two surveys, the number of farmers in this group declined to 0.4 percent.

Finally, there was the group of "absentee landowners" who made up the remaining 3 percent. This consisted of a small share of rural households that were not employed in agriculture at all and leased all their land out to farmers. They were, for example, industry and public services wage employees or pensioners.

Determinants of the Growth of Individual Farming

Mathijs and Swinnen (1998) explain how the shift to individual farming is determined by collective farm members comparing the present value of the expected income from using their assets in an individual farm with the present value of the expected income from alternative uses (for example, in a cooperative farm) net of the exit costs. Exit costs are defined as the costs involved in shifting production factors from the collective to an individual farm. The notion of exit costs is expanded here to that of "switching costs," following Williamson (1985). Switching costs include all the transaction costs a member faces when he or she wants either to leave the cooperative farm or to enlarge an existing private plot. Examples of these costs are investment costs, the costs of searching for factors of production, and the costs of learning new skills.

Based on an intercountry aggregate data analysis, Mathijs and Swinnen (1998) conclude that the shift to individual farming is stronger where: more of the land was distributed to farmworkers rather than restituted to former owners, the share of agriculture in employment is high, and the exit costs involved in withdrawing assets from collective farms are low. To expand briefly the point on the share of agriculture in employment, this share in CEECs is positively related to the individualization of farming for two reasons: firstly, the share of agriculture in total employment is typically negatively correlated with the level of development and the use of technology in agriculture. A country with a high share of agriculture in employment tends to be less developed and uses inferior technology compared with a country with a low share of agriculture, where there is a substitution of capital for labor as a result of the technological progress. Secondly, a high share of agriculture in total employment is an indication that relatively few people have left agriculture. Recent World Bank surveys suggest that the overwhelming majority of individual farmers in CEECs are former cooperative members or employees of state farms (Csaki and Lerman 1997). Since, in general, outsiders lack the appropriate skills to start up a private farm, the number of farmworkers is a good reflection of the number of people actually interested in individual farming.

It is remarkable to see how two countries at the extremes of the spectrum are exactly opposite in the three factors outlined above.

Albania, where the share of individual farms in TAL is one of the highest (94 percent in 1995), distributed all the land to farmworkers, has a high share of agriculture in employment (around 50 percent), and has relatively low exit costs due to favorable government regulations. Slovakia, where the share of individual farms in TAL is lowest (8 percent in 1998), restituted land to former owners, has a low share of agriculture in employment (around 8 percent), and government regulations make exit difficult.

In order to explain the differences between Romanian regions and households in the shift to individual farming, Rizov et al. (2001) use data from the 1996 farm household survey to analyze the determinants of the shift to individual farming from a household perspective. They show that several characteristics of the households determine this organization of production and, more specifically, how land is allocated to individual farming. Three groups of important factors are: (1) the management capability of the household; (2) the household's capital and finance constraints; and (3) the natural, socio-economic and market environment affecting the decision to shift assets to individual farming.

Some of the key insights of the study are summarized here (further details and technical explanations can be found by reference to the study).

The first result was that human capital played a key role in the decision of a member to leave a collective farm and start up an individual farm, because the individual farmer needed to combine a broad range of skills and experiences. A higher standard of education of the household head increased the likelihood of a household shifting to individual farming. However, as the years of education increase, so does the opportunity cost of labor. At some point, other employment opportunities start to dominate the attractiveness of individual farming. For the sample analyzed, this occurred at about eight years, that is, when secondary school was completed. After this level of education, an individual was more likely to opt out of full-time farming in favor of off-farm employment. Consequently, the likelihood of shifting assets into individual farming decreased.

Second, younger farmers were more likely to use their land in an individual enterprise since typically they were more dynamic and more likely to take risks.

Third, individuals who commuted for work from rural areas to town in the pre-reform period were more likely to start individual farming, possibly because they possessed better entrepreneurial attributes or had better contacts with the business environment. These individuals were also likely to possess previous experience with farming as they had always lived in the rural community.

Fourth, the ability of the household to shift assets into an individual farm was strongly affected by its access to the specific assets required for the type of farming undertaken. The study shows that, indeed, ownership of machinery, buildings, or livestock, all had a positive impact on the likelihood that a household in Romania would choose to operate an individual farm. The reason for this is that transition agriculture is characterized by market imperfections, especially for capital and land. Capital constraints are often a serious limitation for potential individual farmers. With capital and land market imperfections present, ownership of capital assets and land facilitates a farmer's direct access to these production factors. Furthermore, it may also improve access to external financial resources.

Fifth, the presence of alternative sources of income within the household, in the form of wages and pensions, positively influenced the decision to farm individually. The external income provided income smoothing and insurance, and an external source of capital to help the household overcome some of the hurdles it faced in initiating individual farming. Access to nonfarm capital sources, such as income from off-farm employment, pensions, or movable assets, may be more effective in securing external finance than farm buildings or land, as financial institutions prefer liquid assets for supporting financial contracts as opposed to illiquid assets such as land. This is especially true in transitional economies where markets for real assets are often highly distorted, nontransparent, or even nonexistent (Swinnen and Gow 1999).

Sixth, security of land tenure, measured by possession of ownership titles, had a positive effect on individual farming. This result confirms arguments made by many experts that implementing full land reform is of crucial importance in order to stimulate more efficient farming.

Seventh, the available labor in the household also had a significant positive impact on the share of own land used by the household itself. This suggests that not only capital constraints but also labor constraints affect individual farming. The study shows that, on average, one additional household member of working age resulted in a more than 15 percent increase in the share of own land farmed individually.

Eighth, factors such as the general social and economic conditions, how conducive the local culture was for entrepreneurship, existing infrastructure, and the distance to urban markets also affected the relative costs and benefits of starting up an individual farm. Therefore, these factors also affect the likelihood of individuals shifting their assets out of collective farms into individual farming operations.

Availability and access to markets and services were important for the development of individual farming. Good access to markets had a positive and highly significant impact on individual farming. This indicates that without access to such markets, individual farming is less likely to develop. Likewise, the availability of inputs such as fertilizers, machinery parts, and machinery services all had a positive effect on individual farming.

Finally, regional characteristics had a highly significant impact as well. For example, more land remained in cooperative structures when the region was dominated by an older population. Not surprisingly, individual farming was more widespread in regions where it was already more important before the reforms. This can be seen from table 3.2. In counties where individual farms already used more than 20 percent of TAL before the reforms, individual farming in 1993 accounted for more than 90 percent on average. By contrast, in regions where very little individual farming occurred before 1989, less than 30 percent of TAL was used by individual farms in 1993. There are several reasons for this, including differences in local culture, social capital, agricultural systems, and regional infrastructure.

Implications

The development of individual farming in Romania is strongly influenced by human capital, physical capital endowments and the access to other financial sources, as well as by the market environment. Human capital characteristics such as age, education, experience, and so on reflect the entrepreneurial input and managerial capacity of the household to manage an independent enterprise.

However, in conditions of severe capital constraints, such as have prevailed in Romania during transition, individuals with sufficient managerial capability might lack sufficient or suitable physical capital to start up an individual farm. The result is that potential entrepreneurs often find it difficult to secure external finance, which prevents them from farming individually within the Romanian socio-economic environment.

Imperfect land and capital markets are thus a very important constraint on the development of individual farming. Direct access to capital goods (machinery, buildings, and livestock) and secure land tenure have a significant positive effect on the shift to individual farming. Furthermore, access to external income sources, such as pensions and wages, stimulates individual farming, as it provides income smoothing, insurance, and a capital source for investment and working capital. For similar reasons, access to inputs, such as fertilizers and machinery parts, encourages individual farming.

In order to stimulate individual farming and private small-scale entrepreneurship, the priority of the Romanian government should be the improvement of the functioning of factor markets in order to reduce the constraints discussed above. Further, investment in human capital (education, training, etc.) is important, not only for agriculture but also for overall rural development. Relaxing the constraints identified here would facilitate the growth of individual farms into commercially viable production units, and the creation of a more balanced agricultural sector production structure. Thus, in the long run, the development of rural markets generating sufficient rural incomes will in turn support the rural location of other private income-generating activities and lead to the sustainable rural economy necessary for accession to the EU.

Acknowledgments

The authors acknowledge the financial support by the PHARE ACE programme of the European Commission (P98-1517-F and P96-6090-R), and thank Zvi Lerman and the editors for their comments on the chapter.

Notes

1. Some sources report 12 percent (European Commission, DGAgri 1998), while others report 9.5 percent (National Commission for Statistics 1986).

References

Carter, M. R. 1987. Risk sharing and incentives in the decollectivization of agriculture. *Oxford Economic Papers* 39:577–595.

Csaki, C., and Lerman, Z. 1997. Land reform and farm restructuring in East Central Europe and CIS in the 1990s: Expectations and achievements in the first five years. *European Review of Agricultural Economics* 24:428–452.

Deininger, K. 1995. Collective agricultural production: A solution for transition economies? *World Development* 23:1317–1334.

European Commission, DGAgri. 1998. *Agricultural Situation and Prospects in the Central and Eastern European Countries: Romania.* Working document, Brussels, Belgium.

Jackson, M. 1997. Political economy of agricultural reform in Romania. In *Political Economy of Agrarian Reform in Central and Eastern Europe*, ed. J. F. M. Swinnen, Aldershot, UK: Ashgate.

Lin, J. Y. 1988. The household responsibility system in China's agricultural reform: a theoretical and empirical study. *Economic Development and Cultural Change* 36:199–224.

Mathijs, E., and Swinnen, J. F. M. 1998. The economics of agricultural decollectivization in East Central Europe and the Former Soviet Union. *Economic Development and Cultural Change* 47:1–26.

Ministry of Agriculture and Food, World Bank, and the European Commission. 1997. *Private Agriculture in Romania: Farm Survey.* Bucharest, Romania. Mimeo-graphed.

National Commission for Statistics. 1986. *Statistical Yearbook.* Bucharest, Romania: National Commission for Statistics.

———. 1994. *Statistical Yearbook*, Bucharest, Romania.

Policy Research Group (PRG), Katholieke Universiteit Leuven. 1999. *Romanian Farm Household Survey 1998.* Leuven, Belgium. Mimeo-graphed.

Pryor, F. 1992. *The Red and the Green. The Rise and the Fall of Collectivized Agriculture in Marxist Regimes.* Princeton, N.J.: Princeton University Press.

Putterman, L. 1985. Extrinsic versus intrinsic problems of agricultural cooperation: anti-incentivism in Tanzania and China. *Journal of Development Studies* 21:175–204.

Rizov, M., Gavrilescu, D., Gow, H., Mathijs, E., and Swinnen, J. 2001. Transition and enterprise restructuring: the development of individual farming in Romania. *World Development* 29:1257–1274.

Rizov, M., and. Swinnen, J. 2002. Post-Communist Agricultural Transformation and the Role of Human Capital: Evidence from Romania, *PRG Working Papers*, No. 22, Department of Agricultural and Environmental Economics, Katholieke Universiteit Leuven, Belgium.

Schmitt, G. 1991. Why is the agriculture of advanced Western countries still organized by family farms? Will this continue to be so in the future? *European Review of Agricultural Economics* 18:443–458.

Swinnen, J. F. M., and Gow, H. 1999. Agricultural credit problems and policies during the transition to a market economy in Central and Eastern Europe. *Food Policy* 24:21–47.

Williamson, O. 1985. *The Economic Institutions of Capitalism.* New York, US: Free Press.

Chapter 4

Rural Credit for Individual Farmers: A Survey Analysis of Rural Financial Markets

Junior R. Davis, Gertrud Buchenrieder, and Kenneth J. Thomson

The macrofinancial situation in any country has important implications for rural finance, while the impact of rural finance on the overall financial situation of a country depends on the importance of the rural sector in the overall economy. In Romania, both relations are significant and, in fact, the second may be the more important (Vogel and Adams 1996).

Heidhues, Davis, and Schrieder (1998) have demonstrated that Romania's financial market could potentially supply the required services to the rural economy, particularly the agricultural sector, without becoming a major burden for the government's budget. Romania's financial depth in terms of its monetarization given by M2[1]/GDP (Tietmeyer 1994, 421–433) is satisfactory, and the density of rural banking outlets available per 10,000 persons is quite favorable. However, Romania's rural economy, particularly the private agricultural sector, has a huge unsatisfied effective and latent demand for financial services, especially credit (see Heidhues, Davis, and Schrieder 1998). Market deficiencies with regard to the variety, quality, and efficiency of financial services offered in rural areas prevent private agriculture from acquiring the necessary momentum in its economic transformation.

In this chapter, the main determinants of the financial problems are identified by utilizing the results of a representative survey of individual farms and farmers' associations in Romania. The aim is also to analyze, from the perspective of small, private (largely

- an income generation module—factor and nonfactor income data;
- an asset module—consumer and producer asset balance and variability;
- a labor supply module—family and nonfamily labor; and
- a financial intermediation module—debt, credit, and savings balance and variability.

Separate questionnaires were applied to the demography of the farm households and to the "enabling environment" of the communities (Davis and Hare 1997). Besides the microeconomic data at the farm level, secondary data on the formal financial market, provided by the National Bank of Romania, and first-hand information from commercial bank managers, were also integrated into the analysis. The microeconomic data for the dynamic and non-homogeneous private farm sector, for which official statistics usually provide scarce and outdated data, was a particularly relevant contribution since it provided detailed information.

Specific Problems of Rural Finance in Romania

The main rural financial problems facing Romanian agriculture may be categorized into three groups: debt problems, operational problems, and farm restructuring problems in an "enabling" environment dominated by government intervention.

Debt problems concern the existence of debts with state integrators (state downstream companies), local banks, and moneylenders. Interenterprise arrears accumulated prior to the transition process are a less-important problem for small private farms. Higher-than-expected inflation and high nominal interest rates have exacerbated the debt problem and made it difficult for farmers to service loan repayments out of earnings. Fifty-nine percent of the survey respondents did not apply for a formal loan because of high interest rates; five percent did not have sufficient collateral.

Operational problems mainly relate to financial issues concerning inadequate cash flow and profits to cover operating and investment expenses. These problems may derive from a variety of sources, for instance: input price rises higher than anticipated interest rates; a lack of financial management tools such as deposit instruments to manage the flow of internal savings of the farm; and inexperienced farm management of cash-flow fluctuations, often a function of production seasonality. The results of the survey show that the return on capital for Romanian farmers is often lower than the interest rate charged (Davis, Gaburici, and Hare 1998).

Farm restructuring problems mainly concern the "enabling environment." As in most CEECs, agricultural investment in Romania has fallen. This may not be the result of increased interest rates, but rather due mainly to broader farm restructuring problems and macroeconomic instability, as well as to the slow pace of land reform and agri-food industry restructuring (Davis 1997).

An important aspect of the process of creating a viable rural financial market in Romania is the development of a legal framework for landownership, and land and lease markets. This will enable land use to reflect commercial decisions supported by land as collateral for loans. Whatever the mechanism adopted for initial privatization, the creation of viable, freely functioning land markets is essential for providing a flexible mechanism for reorganization and improved resource allocation.

The survey also sought to identify the chief problems expressed by individual farmers and FAs regarding their interaction with financial intermediaries and running their farms. According to the data analysis, these were primarily the lack of marketing and processing facilities for agricultural outputs. Liquidity constraints were ranked third among the farmers' problems (see Heidhues et al. 1998).

Interaction of Farmers and the Rural Financial System: Liquidity Constraints

One of the major objectives of the farm survey involved the identification of the main reasons that determined the low level of demand for loans. In the first survey questionnaire, four pre-established reasons were included, but the enumerators were instructed to note any other reasons given by the farmers. A comparative analysis of the data shows a significant change for each reason as a share of the total number of responses between the first and second survey rounds. In the first round (April 1997), the dominant reason given was "Could not fulfil the loan conditions," whereas in the second survey (in December 1997), the main reason was that the "Interest rate is too high" (see table 4.1). This change may be explained by the fact that during 1997 the credit conditions imposed on private-sector borrowers were relaxed, and a greater diversity of credit sources became available. At the same time, the government considerably reduced the subsidization of interest rates for farm loans, but introduced some assistance through the voucher scheme described in chapters 1 and 2. As 1997 ended with an inflation rate of 152 percent, the farmers' fear of high interest rates was probably justified. Certainly, it was mainly large ASs or CCSCs that benefited the most from government subsidized interest rates and subsidy programs (Davis and Hare 1997).

To better understand how individual farmers and FAs perceive the interest rate, the following question was introduced in the second survey questionnaire: "What is the maximum annual interest rate acceptable to you?" Around 50 percent of the farmers in the sample answered this question. An interest rate of 60–75 percent, which was considered acceptable by 21 percent of the farmers, is in fact similar to the bank lending rates of, for example, the Credit Union (Co-operative de Credit [CC] or Popular Bank) in 1997. This interest rate would also have been acceptable to commercial banks for 1998, if the inflation rate had not exceeded the levels forecast for the 1998 budget (42 percent). However, the majority of farmers (68 percent) indicated that an interest rate of 5–20 percent per annum, some of the respondents emphasizing "subsidized interest rates," would be more acceptable (Davis and Gaburici 1999). The farmers' desiderata concerning interest rates and the term of future loans were quite unrealistic, given that in 1996 farmers were paying a subsidized interest rate of 39 percent (Davis and Hare 1997).

As the majority of farmers in the first survey round maintained that they did not apply for loans because they could not meet the bank requirements to obtain credit, the second questionnaire was enlarged to consider this issue more closely. The results of the "sub-reasons" for failing to apply for a loan showed that the main condition that could not be met was that of farm solvency, or sufficient income.

Table 4.1 Main Reasons for Not Applying for a Loan, December 1997

	County							
	Brasov		Dolj		Timis		Total	
	No.	%	No.	%	No.	%	No.	%
1. Could not fulfil the loan conditions	16	24.6	9	14.3	2	3.4	27	14.5
2. Procedure too complicated	3	4.6	3	4.8	2	3.4	8	4.3
3. No need, enough private money	7	10.8	18	28.6	13	22.4	38	20.4
4. No need, current loan	4	6.2	8	12.7			12	6.5
5. Afraid of going to an organization	1	1.5			1	1.7	2	1.1
6. Interest rate too high	13	20.0	14	22.2	32	55.2	59	31.7
7. No confidence in credit organization			3	4.8	2	3.4	5	2.7
8. Did not like to have credit	11	16.9	4	6.3	3	5.2	18	9.7
9. Other	10	15.4	4	6.3	3	5.2	17	9.1
Total	65	100.0	63	100.0	58	100.0	186	100.0

Source: Romanian rural finance survey 1997/1998; own data

The loan conditions faced by the farmers are applied by most commercial banks. Only the loan-making procedure of the CC is different, as it favors its members and those who have previously deposited at a cooperative or a mutual trust fund. Its loan conditions require the applicant to state the volume of his or her deposited funds with the cooperative and the size of loan requested. In the CC-applicant contract, the annual interest rate, the risk commission (which is collected from the debtor as soon as the loan is approved), and interest rate penalties (which are collected over the current interest rate in case of default) are stipulated in percentage terms. Loan contracts also typically contain a clause allowing the CC to revise (with notification) the loan interest rate as a result of a change in the rate of inflation, or to refinance the interest rate. The CC asks the debtor for his or her own assets as guarantees, and also for part of the guarantors' assets.

In the case of the Banca Agricolă and other commercial banks, the loan contract stipulates a fluctuating interest rate established by the bank, as a function of the costs of its credit sources. If the borrower delays debt repayments, he or she will pay an increased interest rate. Between the date of loan approval and its effective use, the borrower is obliged to pay to the bank an annual commission for nonutilization. A further condition imposed by Banca Agricolă requires the presentation of a land "ownership title," certifying the farmer's full property rights to the land. Owing to the slow progress with the administration and implementation of land titling procedures, the receipt of these titles is usually subject to great delay. The second survey round found that 7 percent of those farms that could not meet the loan requirements blamed this on the lack of an "ownership title."

During the second survey round, farmers were asked to estimate the capital requirements for their farm (table 4.2). Few farmers reported that they had borrowed money in 1997, at least through formal channels. However, 55 percent of the sample provided an estimate of capital requirements for 1997. The average requirement per farm was $8,984, while the average private farm sales revenue for 1997 was $2,620 (Davis and Gaburici 1999). Farmers thus estimated the capital requirements, or required level of borrowing, for 1997 at roughly three times the current level of sales, which appears highly exaggerated. These estimates may be attributed to inexperience, which leads to the inability of farmers to forecast their financial needs correctly.

The main characteristics of the credit demand side in the small-scale private agricultural sector are risk aversion and, to some extent, traditional sociocultural behavior with regard to external finance. The investments of the individual farmers and FAs in the private agricultural sector can be financed through relatively small loans. In general, individuals demonstrate much better credit repayment behavior than the CCSCs. Empirical evidence shows that individual farmers repay 75 percent, FAs, 47 percent, and CCSCs, 26 percent, of their debt to the formal financial sector on time (Heidhues and Schrieder 1997). Nevertheless, at the level of the individual agricultural entrepreneur and the lender, factors other than the price of credit result in credit market access constraints. Latent borrowers do not voluntarily enter the credit market due to reasons that are not related to price (Davis, Gaburici, and Hare, 1998).

Uncertainty, screening problems, and adverse selection problems with respect to loan repayment and collateral values dominate the behavior of the financial intermediaries in Romania toward small farm borrowers. These intermediaries prefer their traditional clients, the CCSCs, although loan loss is the highest in this market segment of the agricultural sector. The intermediaries offer moderate interest rates, grant large but few loans, and require securities that represent 200 percent of the loan principal. In the survey, loans to the private agricultural sector were funded primarily out of targeted and subsidized loan funds provided by the government. But, even here, the CCSCs were the main beneficiaries, receiving 98 percent of the credit funds. With the exception of the Credit Union, formal financial intermediaries in Romania are very reluctant to finance private agricultural investments out of their own loan portfolios. Consequently, credit supply and demand in the private agricultural sector do not match.

Conclusion

On the supply side, small individual farmers in Romania need improved access to sources of rural finance if they are to operate more efficiently and interact more effectively within the agri-food system. To facilitate this, further agri-food de-monopolization and

Table 4.2 Capital Requirements as Estimated by Romanian Farmers, 1997

		County			
		Brasov	Dolj	Timis	Total
Circulating Capital	No.	25	10	2	37
	No. as % of sample	31.3	13.0	3.2	16.8
	Sum ($)	93,045	8,001	1,484	102,530
Fixed Capital	No.	41	36	30	107
	No. as % of sample	51.2	46.8	47.6	48.6
	Sum ($)	164,667	201,189	591,695	957,552
Total Capital	No.	49	38	31	118
	No. as % of sample	61.3	49.4	49.2	53.6
	Sum ($)	257,713	209,190	593,179	1,060,082
Sample		80	77	63	220

Source: Romanian rural finance survey 1997/98; own data

privatization should be encouraged by the government. Central, in terms of releasing working capital for farmers, should be the reform of the practice by which state silos store rather than purchase grain from farmers. In charging rental fees for storage, they both deny farmers working capital and also pass on the risk associated with storage. Similarly, food input-output price disparities need to be addressed, but not through government-mandated lending programs, which are expensive and have exacerbated inflationary pressures in Romania. Perhaps a more targeted and—in terms of outreach—more limited scheme, devised along purely commercial or minimum-collateral lines aimed at emergent farmers, would be more sustainable and effective (Davis and Hare 1997).

Microcredit schemes have been reasonably successful in other transitional and developing countries in terms of facilitating access to credit and encouraging savings mobilization for the rural poor and small enterprises (OECD 1996). As in other CEECs in recent years, such schemes, although mainly aimed at nonrural small- and medium-size enterprises, could be more widely and effectively utilized in Romania. Similarly, a comprehensive rural development program would encourage the development of more nonfarm income-generating enterprises, breaking the cycle of subsistence and poverty in which most small farmers have found themselves.

On the demand side, individual farmers and FAs often expressed their need for credit to undertake capital investments, yet low cash rates of return and high uncertainty regarding the ability to service a loan led to a low level of effective demand for borrowed funds (Davis and Gaburici 1999; Safavian and Meyer 1999). Romanian private farms are generally undercapitalized and in great need of working capital, particularly for the purchase of farm inputs. However, relatively few of the survey respondents actually applied for formal credit. The main reasons for this were that farmers felt that they could not fulfil the loan conditions or that the interest rates were too high. Many of the impediments to farm investment thus occur at the farm and production level, rather than at the level of the financial systems. These impediments include a lack of economic opportunities, low nonfarm and farming incomes, and limited investment opportunities, which may present more significant deterrents to farm investment than the supply-side barriers discussed previously (Safavian and Meyer 1999).

Nonetheless, the building of an efficient and effective rural financial market in Romania requires a multifaceted approach. An essential requirement is financial discipline and monetary stability, which cannot be achieved if the Central Bank, is under government command and is pushed into financing loss-making state enterprises (Heidhues, Davis, and Schrieder 1998). The government should not interfere with the Central Bank, whose key mandate should be securing monetary stability, a precondition for effective financial market development. In rural areas, even if institutional density is well developed, as is the case in Romania, financial innovations are required at four levels to improve the access of private agricultural and nonagricultural enterprises to the financial market:

1. at the financial system level, creating a reliable, fair, and enforceable regulatory framework with an effective supervisory structure;
2. at the organizational level, either restructuring banking institutions to better serve the rural client market, or strengthening existing rural finance institutions, such as the credit cooperatives in Romania, or instituting new ones if a reform of existing structures turns out to be infeasible;
3. within finance institutions, streamlining application, approval and supervision processes, and integrating participatory client involvement; and
4. offering new services geared to the needs of the rural clientele.

These are innovations that require a major long-term effort in education, training, gathering experience, changing laws, and restructuring other institutions and organizations. This is a process requiring time, patience, and persistence; a shock therapy approach will not succeed.

Notes

1. M2 comprises cash in circulation and retail deposits at banks and building societies.

2. The administrative organization of the Romanian territory comprises forty-one counties. All the official statistical data are reported at the county level. The administrative system was adopted for the sampling procedure, which has the advantage of being a basis for comparison between the results of the present analysis.

References

Davis, J. R. 1997. Agricultural reform during transition: a synopsis of recent developments and future prospects. *MOCT-MOST: Economic Policy in Transitional Economies* 7:(4) 55–79.

Davis, J. R. and Hare, P. 1997. Reforming the systems of rural finance provision in Romania: some options for privatization and change. *Quarterly Journal of International Agriculture* 36:(3).

Davis, J. R. and Gaburici, A. 1999. Rural finance and private farming in Romania. *Europe-Asia Studies* 51:(5) 843–869.

Davis, J. R., Gaburici, A., and Hare, P. G. 1998. What's wrong with Romanian rural finance? Understanding the determinants of private farmers' access to credit. *CERT Discussion Paper No. 98/8*. Edinburgh: Department of Economics, Heriot-Watt University.

Heidhues, F. and Schrieder, G. 1997. Access constraints of Romanian peasants in relation to the formal financial sector. *CERT Discussion Paper No. 97/12.*

Heidhues, F., Davis, J. R., and Schrieder, G. 1998. Agricultural transformation and implications for designing rural financial policies in Romania. *European Review of Agricultural Economics* 25:(3) 351–372.

Heidhues, F., Schrieder, G., Hare, P.G., Davis, J.R., and Gaburici, A. 1998. Empirical research on the financial market in Romanian private farm sector. *Romanian Economic Research Observer*, nos. 3–4. Bucharest, Romania : Center of Economic Information and Documentation.

OECD. 1996. *Microcredit in Transition Economies.* Paris: OECD/GD (96)40.

Safavian, M. and Meyer, R. L. 1999. Obstacles to the provision of finance to the private farm sector in Romania. *IX European Congress of Agricultural Economists.* Organized session papers, Warsaw, Poland: 186–201.

Tietmeyer, H. 1994. Deutsche Bundesbank. In *Handwörterbuch des Bank- und Finanzwesens*, (Hrsg.) eds W. Gerke und M. Steiner 421–433. Stuttgart: Schäffer-Poeschel Verlag.

Vogel, R. C. and Adams, D. W. 1996. *Rural Finance in Romania.* Columbus, OH: Ohio State University, Rural Finance Group. Draft Report.

Chapter 5

Credit for Emerging Commercial Farmers: A Survey Analysis

Christine Coulomb and Sophia Davidova

This study discusses the access of Romanian commercial farmers to credit, and complements the previous chapter where the credit market for small private farmers was assessed. Since little or no internal accumulation has taken place during the years of transition, most farmers lack the equity to invest in farming, and, therefore, need external sources of funding. Credit would appear to be a practical way to provide such funds, but recent surveys show that Romanian farmers use very little credit from banks. The study investigates the reasons for the lack of credit flows between banks and farmers or, in other words, analyzes the constraints to farm credit. Constraints that induce a lower demand for credit are referred to as internal constraints, and constraints which impede farmers' access to credit are defined as external constraints.

The study is focused on two types of farm structure in Romania, both of which are private and commercially oriented: commercial companies with private capital (CCPCs) and agricultural societies (ASs). It is based on a survey that took place in summer 2000. It incorporates only farms cultivating more than 100 hectares of cereals and sunflower in the Câmpia Romàna (Romanian Plain), namely the counties of Ialomiţa and Teleorman. They are located on the Danube plain in the south of Romania. Slobozia, the main town of the Ialomiţa county is 120 km east of Bucharest, while Alexandria, the main town of the Teleorman county, is 90 km southwest of Bucharest and 40 km north of the Bulgarian border. When the term "farmer" is used in this study, it, therefore, means "the farm manager of a larger cereal farm." Thirty farmers were interviewed, but one outlier was removed and thus twenty-nine cases were analyzed. Bankers involved in agricultural credit were also interviewed. Whenever possible, the people interviewed were bank employees in charge of loans to farmers, rather than managers of the bank branches.

Institutional Framework of the Credit Market in Romania

Interest rates are generally considered to be one of the most important variables in the macroeconomic setting and a good proxy measure for the soundness of financial markets. The criteria most scrutinized by analysts are the absolute level of interest rates as well as relative rates; real deposit rates being expected to be above the real GDP growth. One important comparison concerns the evolution of various interest rates. Figure 5.1 shows the evolution of the banks' deposit and lending rates, and of the consumer price index (CPI) over the period March 1998–May 2000. This comparison displays the nominal levels of these interest rates, all three being above 40 percent for nearly the whole period. It also shows the relative variation of interest rates. The lending rate is generally expected to exceed both the deposit rate and inflation and that has been the case since September 1998, but not before. The rate on deposits is also expected to exceed inflation but, in general, this has not happened. Between September 1999 and May 2000 real deposit rates were negative. This can be seen in figure 5.1, where the dotted line, representing the deposit rate, is below the CPI, which is an indicator of the general inflation.

Heidhues and Schrieder (1997) analyze the performance of the formal financial market in terms of its "financial deepening" and they note that the monetarization of the economy, as measured by the ratio of M2—cash in circulation plus retail deposits—to GDP, is a good indicator of the financial system's efficiency in mobilizing funds. The indicator M2/GDP shows that Romania's financial depth is far less than would be expected for a country with that level of GDP. After a rapid increase from 16 percent in 1992 to 25 percent in

1995, the ratio seems to be leveling off. By contrast, in Hungary for example, M2/GDP was 46 percent by the end of 2000.

Figure 5.1 Monthly Development of Bank Deposit and Lending Rates, Discount Rate and CPI, March 1998–June 2000

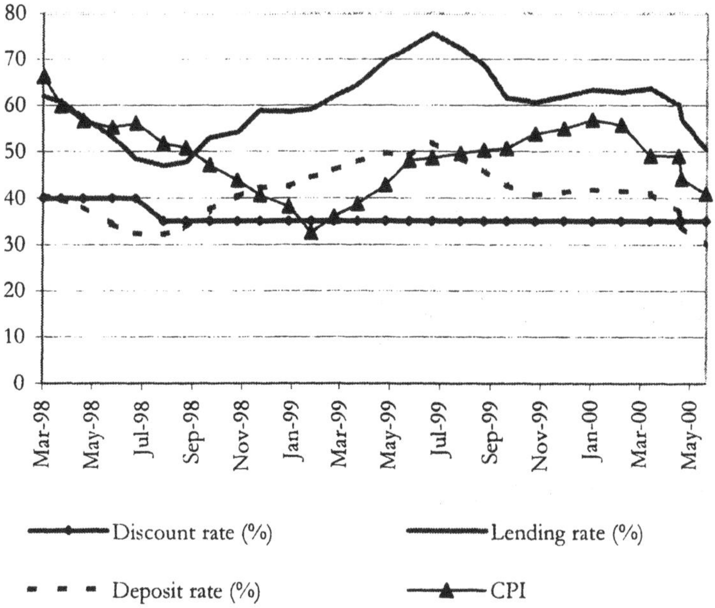

Source: National Bank of Romania (1999, 2000)

The financial market in Romania also appears to be repressed. McKinnon (1991) explains how governments may distort their domestic capital markets to such an extent that the economy becomes "financially repressed." Restrictions usually include a cap on the authorized rate of interest to be charged to customers, reserve requirements on bank deposits, and compulsory credit allocations. McKinnon notes that such an unfriendly institutional framework may be aggravated by inflation. The institutional framework of the credit markets in Romania seems to display most of the above-mentioned characteristics. High inflation, reserve requirements, and commanded credit allocation suggest the existence of repression in the Romanian banking system. Reserve rates imposed by the National Bank of Romania on lei bank deposits were increased stepwise from 12.5 percent in March 1998 to 30 percent since December 1999. This may have important consequences for the profitability of the commercial banks' lending portfolio, since the banks can only lend the amount of deposit net of mandatory reserves. The higher the rate of reserves and the lower the compensatory rate paid by the central bank, the higher the bank needs to raise its lending rate to cover the opportunity cost of mandatory reserves. Among the adverse consequences of this situation, two are particularly important for farmers' access to credit. One is the reduction of the flow of loanable funds through the organized banking system. The other is the arbitrary variation in the interest rate charged to different classes of borrowers, depending on whether or not they are favored by the bankers.

Another institutional problem is raised by the banking system itself. Heidhues and Schrieder (1997) studied the number of formal financial organizations per 10,000 persons, and the range and quality of services provided. In 1997, there were about 1.3 financial intermediaries per 10,000 persons employed in agriculture, and the authors concluded that the rural financial institution network in terms of banking outlets was relatively well developed. The list of Romanian banks and their rural outlets is shown in table 5.1.

Table 5.1 Romanian Banking Enterprises and Their Rural Outlets, March 1997

Bank	Branch	Subbranches		Agencies	
		Total	Rural	Total	Rural
Banca Agricolă (BA)	41	188	59	78	65
Banca de Credit Cooperatist (Bancoop)	44	178	51	5	6
Banca Comercial a Romana (BCRO)	41	123	15	48	21
Banca Renasterea Creditului Roman esc (Credit Bank)	49			67	12
Banc Post (BPST)	45	22	1	38	6
Fourteen other banks	154	29	2	22	2
Total	418	608	137	313	146

Source: Heidhues and Schrieder (1997)

However, the banking system has undergone drastic changes since 1997. The Romanian Development Bank (BRD) and Banc Post have been privatized. Bancoop, Banca Albina, and Credit Bank have suspended their operations. Banca Agricolă was being reorganized to prepare for privatization and was finally privatized in 2001. A new bank, Banca Internationala a Religiilor (BIR), was created in 1995, but in 2000 went into bankruptcy. Thus, the sources of supply of loans to agriculture seemed to be drying up by 2000.

Collateral is another impediment to farmers' access to credit. Farmers experience problems in providing sufficient assets as collateral for bank loans. The use of farmland as collateral in Romania is very low. In the case of ASs, land is not part of their assets. The members could decide in a general meeting to pledge their land, but this does not appear to be the case. Moreover, until late 1998 Romanian law prohibited transactions with restituted land. Therefore, this land could not be used as collateral. There was some progress in 1999–2000, but the land market is still quite thin because the structural weaknesses of land ownership, fragmentation of ownership, incomplete titling, and poor registration, impede the use of land as collateral. Only a few banks seem to accept future harvest as collateral. The high collateral to disbursement ratios imposed by banks, together with their prudence regarding the choice of assets, is in sharp contrast to the practice of the nonbank credit sector, which operates without any collateral requirement.

Therefore, the institutional framework in Romania is not favorable to agricultural credit. It is not surprising that surveys conducted by the Institute of Agricultural Economics in Bucharest (Gavrilescu and Sarris 1996), by the World Bank (1996), and by the MoAF/World Bank/EU (1997) point to the small reliance of private farmers on bank credit.

Methodology

The study is based mainly on primary data. Both farmers and bankers were interviewed. Data were collected concerning information on the farmers' use of variable inputs, a proxy that was chosen to measure a possible lack of credit. Data were also collected about farmers' past and present borrowings. Farmers were asked about their expectations on yields, prices, and production costs; the price they paid for credit obtained from both banks and input suppliers; and about their decision-making procedure.

Thirty farmers were interviewed, but one outlier was omitted in the subsequent analysis. The survey took place in the counties of Teleorman and Ialomiţa. These counties were chosen because they are situated on the Danube plain, a region with large areas of sunflower and wheat cultivation.

The criteria for selection were:

- farms with a legal structure organized either as an AS, or as a CCPC;
- farms cultivating between 100 and 1,500 hectares of mostly, if not exclusively, cereals, either without livestock, or with livestock only as an ancillary activity;
- cereal farms whose main crops in summer 2000 were wheat, sunflower, and maize;
- farms which were in touch with the regional offices of the extension service agency.

The choice of two crops, wheat and sunflower, was made as these are the main crops on the Danube plain where the interviews took place. A major part of the area in the sample farms, 57 percent in ASs and 64 percent in CCPCs, was under these two crops.

The sample of farms consisted of sixteen ASs and thirteen CCPCs. The latter included both limited liability companies and joint stock companies. Some basic characteristics of the sample farms are presented in table 5.2.

Table 5.2 Average Land Area and Ownership of Land in the Sample Farms, 2000

	ASs	CCPCs
Total land area (ha)	843	519
Land under wheat and sunflower (ha)	479	334
Land owned by the manager (ha)	6	19
Land owned by members and their families (ha)	664	0
Average age of general manager	45	45
Years in formal education of general manager	over 7	over 7
Total number of workers	43	38
Seasonal workers	28	19

ASs are larger according to the land area and they mainly cultivate land owned by their members. CCPCs mainly rent land.

In all cases but two, the respondents were the general managers of the farm. On average, they were forty-five years old, the youngest being twenty-nine and the oldest, sixty-two. With one exception, all had over seven years of schooling, thirteen had gone to university and fifteen to an agricultural school.

The workforce of the farms was predominantly seasonal. In the ASs, out of an average forty-three workers per society, twenty-eight, or almost two thirds, were seasonal. In the CCPCs, which employed thirty-eight people on average, half were permanent workers and half were seasonal. Considering the area cultivated, an average of 843 hectares for ASs and 519 hectares for CCPCs, relatively more people, and particularly relatively more permanent workers, were employed by the CCPCs than the ASs.

The number of members in AS varies from 111 to 1,100, each member bringing on average between 0.4 and 2.8 hectares. This means that, per unit of land cultivated, some societies have seven times more members than others. In addition, both ASs and CCPCs provide machinery services (generally plowing and harvesting) to a relatively large land area belonging to other farmers, 160 hectares on average.

Eight bank employees both at the headquarters of the banks and in local branches were interviewed. Whenever possible, the people interviewed were bank employees in charge of loans to farmers rather than the managers of regional branches. They were working for banks relatively active in agricultural lending: Banca Agricolă (BA), the BRD, Banca Comerciala Romana (BCR) and the savings bank, Casa de Economii şi Consemnaţiuri (CEC).

Analytical Framework

Investigating money issues such as credit is a difficult endeavor because respondents often tend to blur their answers. To ensure more accurate answers, the choice was made to avoid direct questions about money and to use proxies.

First, the quantity of variable inputs used on crops, or, rather, the difference between these quantities and those recommended by the Institute of Agricultural Economics, was chosen as a proxy to assess the farmers' apparent need for credit. Once it was established that farmers use less input than recommended, there were two possible explanations. One was that they have themselves chosen to use less, thus restraining their demand for inputs and, therefore, credit. The other was that they would have liked to use more, but were unable to do so for lack of money. The two situations are different, since they are caused by different types of constraints; internal constraints in the first case, and external in the second case.

Second, the apparent demand for credit (the reported shortage of funds to buy desired inputs) was compared with effective credit requests to the bank, that is, the actual demand for credit. These two possibilities were developed analytically under the assumption that there was no shortage of inputs in the market. This assumption appeared to be realistic as none of the farmers interviewed mentioned that physical shortage of inputs was a problem. Anecdotal evidence suggested that all the main western-produced inputs, seeds, fertilizers, and pesticides, were available along with domestically produced inputs.

In the analysis of external constraints, the first step was to assess the relative importance of lack of credit as a limiting factor of input use, and the second was to try and identify possible causes of credit market failure. Various macroeconomic elements, such as high inflation and mandatory reserve rates imposed by the central bank, might have combined to create an environment where the market for credit was missing. The banker's perception of the quality of the prospective borrower was also very important in the decision-making process. Although the banker would have formed an a priori opinion about the quality of the farmer requesting a loan (prior belief), the farmer might have tried to influence the banker into believing that they belonged to the good risk category (posterior belief). Since the signals that can achieve such a goal are probably situation specific, the influence of several signals was investigated in the Romanian context.

Summary of Results

The Use of Variable Inputs as a Proxy of the Need for Credit

The study of variable input use on wheat and sunflower during the 1999–2000 agricultural year shows that commercial farmers in Romania tended to use less variable inputs than those recommended by the Institute of Agricultural Economics and less than they would have liked. Farmers used less variable inputs than recommended, with the exception of sunflower seeds and phytosanitary products for wheat. The shortfall in input use was particularly important for fertilizers and wheat seeds, since many farmers produced their own seeds rather than buying them. It appears that farmers were quite rationally allocating the funds available for purchase of variable inputs. For example, they mentioned that far better yields were obtained with foreign sunflower hybrids, while they also said that sunflower needed both less fertilizer and phytosanitary treatment than wheat. As a result, farmers bought high-quality seeds and limited their use of fertilizers and phytosanitary products. If they could, however, they would have liked to use far more variable inputs than they did at that time. The main part of the increased expense would have gone on fertilizers.

The lack of credit appeared to be one of the most important factors limiting the use of variable inputs, since twenty-eight out of twenty-nine farmers mentioned it. However, it also appeared that for a number of farmers, lack of irrigation is just as important a factor, if not a more important one. To put these answers in perspective, it is important to note that the interviews took place on the Romanian plain, where lack of water is an endemic problem. Also, summer 2000 was marked by one of the most severe droughts in decades.

Nonbank sources of financing were also investigated and suppliers' credit emerged as the main source of funds for the purchase of variable inputs. This raised a number of issues. The use of suppliers' credit was found to be general, twenty-six out of the twenty-nine farmers requested a credit from their suppliers. However, this credit did not fulfill their financing needs and eighteen of them reported that suppliers had only granted part of the loan requested. Suppliers' credit in the western economies is simply "accounts payable," which means that it does not require special arrangements because its terms are limited (usually three or four months), and it represents only a small part of farm financing for inputs. This is different in Romania. Since bank credit to agriculture is very limited, suppliers' credit has become the main source of financing for many farms. Suppliers, therefore, cannot share their risk with other institutions as is the case in the West. Moreover, the term of this credit is longer than in the western economies, often longer than eight months, that is, the whole cropping season from planting to harvesting. Finally, two issues make it difficult to calculate the cost of suppliers' credit in Romania. The first is that inputs are invoiced to farmers in U.S. dollars, exposing them to currency risk, since farmers receive Romanian lei for their output rather than U.S. dollars. The second is that the high rate of inflation that has prevailed in Romania over the past years makes it very difficult to forecast prices. Farmers find it hard to assess the real cost of suppliers' credit. When asked to forecast this cost over the current period, only six farmers produced a realistic estimate, framing this cost between 25 and 40 percent. Fourteen farmers underestimated it (as equal to or less than 20 percent), and one farmer probably overestimated it (as 100 percent). Furthermore, no farmer could produce a table of amortization of suppliers' credit showing when and how much credit had been undertaken, and when it was due for repayment.

Internal Constraints on the Use of Inputs and Credit

The internal constraints to credit could be explained by the uncertainty farmers face regarding crop yields and output prices combined with the farmers' risk aversion. So, it is to be expected that there would be a relationship between the uncertainty that farmers perceive, their risk aversion, and the amount of variable inputs that they use. Uncertainty faced by Romanian farmers is very high. The range of prices and yields expected by the farmers in the study from their wheat and sunflower crops varied widely from one farmer to another, but it appeared that on average farmers were fairly pessimistic about prices. The yields that farmers expected were always somewhat above the yields that they considered necessary to cover their costs, while the prices that they forecasted were always below the prices that they considered necessary to cover the costs.

Farmers were also asked about their estimated production costs for each crop. ASs systematically mentioned higher production costs than did CCPCs.

Farmers' Self-Imposed Constraints on the Use of Credit

The uncertainty that farmers confronted, particularly concerning prices, partly explains why they used a lower level of variable inputs. However, with only one exception, all farmers mentioned that they would like to use more variable inputs and that they would like to borrow from banks for this purpose. It could be expected, therefore, that they would request loans from banks. But this was not the case.

Figure 5.2 shows a decision tree established following the answers from the farmers. It appears that, whereas twenty-two out of the twenty-nine farmers mentioned that they wanted to borrow money from banks, only sixteen actually requested a loan.

Figure 5.2 Decision Making for Loan Requests (All Farms)

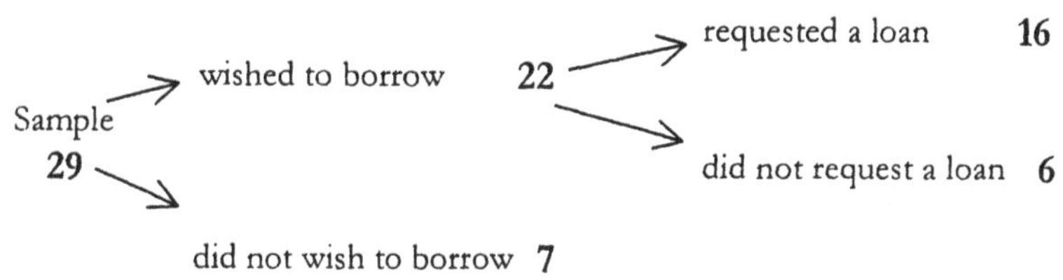

Twenty-seven out of the twenty-nine farmers wished to use more inputs. On average, they wished to use lei 1,213 million; in other words, 639 million more than the 574 million they actually used (i.e., more than twice as much on average). The additional amount of inputs desired appeared to be a useful indicator of the farmers' need for credit. There was an important difference between the additional amount of inputs wished by those who actually requested a loan (lei 789 million) and those who did not (lei 424 million).

Many farmers gave several reasons why they did not wish to borrow or did not request a loan. The answers are given in table 5.3. Cost of credit ranked first as a reason discouraging farmers, but six of them also mentioned that they could not meet the conditions imposed by their bank, while five farmers feared a refusal. Three farmers mentioned that they did not request a loan because their bank had no "credit ceiling," so they perceived that their bank had not been allocated funds to lend to agriculture. It could be that their bank was not allowed to make loans, which was reportedly the case for Banca Agricolă. It could also be that these farmers were confusing ordinary loans, which carried a market interest rate, with subsidized loans. These were a source of cheap credit (18 percent) but were a constrained credit resource, since the related government program was progressively reduced from lei 550 billion in 1997 (about $76 million at the exchange rate at that time) to lei 300 billion ($15.6 million) in 2000.

Table 5.3 Reasons Given by Farmers for Not Wishing to Borrow or Not Requesting a Loan

	Cost of Credit	Could Not Meet Condi- tions	Fear of Refusal	Other Reason	Lack of Credit Ceiling	Did Not Know a Bank
ASs	5	4	3	3	1	0
CCPCs	3	2	2	2	2	1
Total	8	6	5	5	3	1

The pattern was not the same for the ASs and the CCPCs. Figure 5.3 provides a decision tree detailed by type of farm structure. Of the nine CCPCs and thirteen ASs that wished to borrow, only one company did not request a loan, whereas five agricultural societies did not do so. Altogether only eight out of sixteen ASs (50 percent) and eight out of thirteen CCPCs (61 percent) requested a loan. This might be due to the influence of the different process of decision making between commercial companies and agricultural societies. For the ASs, it was the manager who said he wished to borrow, but this opinion might not have been shared by the general assembly of members.

Figure 5.3 Decision Making for Loan Requests Broken Down by Farm Structure

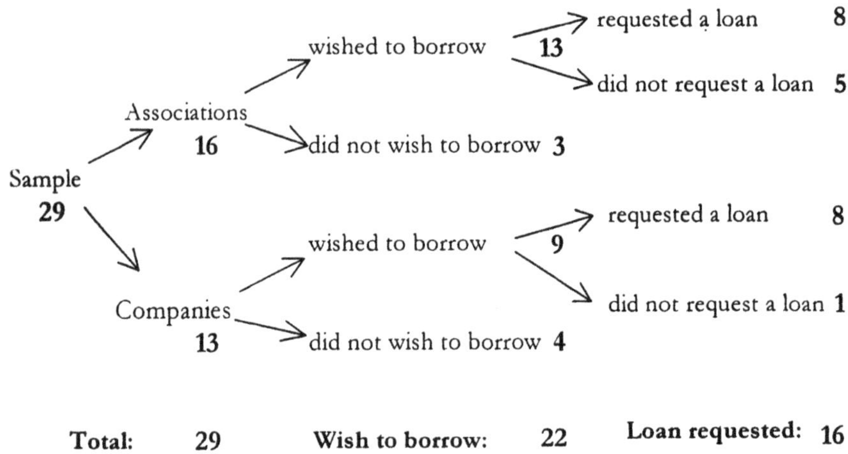

Difference in Production Costs of ASs and CCPCs

In the literature it has often been asserted that, due to a particular type of management, producer cooperatives are inefficient. ASs are in fact producer cooperatives. Their managers reported much higher production costs (yields and prices necessary to cover production costs) than managers of CCPCs. The results are summarized in table 5.4. Whereas the yields and output prices were quite similar for the ASs and the CCPCs in the case of the sunflower, the wheat yields and prices necessary to cover the costs reported by ASs were higher than for CCPCs. More importantly, the production costs estimated by ASs were above those of CCPCs by 32 percent for wheat and 29 percent for sunflower, although all farmers interviewed operated in similar agro-environmental conditions.

A possible explanation could be that, since AS managers are employees rather than owners of the farms, they may be less familiar with financial accounts than are the managers of CCPCs. The latter own the business and are "entrepreneurs," in the western meaning of the term. In this case, the production costs that AS managers quoted could be a ballpark figure. During the interviews, the CCPC managers could readily answer all the questions, whether technical (what is planted where, how it is grown), or financial (input prices), while AS managers had to refer to accountants or technical engineers to provide the information. In this respect, it is interesting to note that the education of AS managers differed from that of CCPC managers. Most of the AS managers were trained in agricultural schools, whereas most of the CCPC managers had attended university. Their prior experience also differed. Two thirds (eleven out of sixteen) of the managers of AS were previously members of the management team of cooperatives, either as a chief engineer or as a head of a farm. In contrast, only three of the thirteen managers of CCPCs had worked in a cooperative before, and most of them came from other professions, such as accounting or trade, or had previously set up their own company. Given their academic and prior work experience, the managers of CCPCs might, therefore, have been more highly trained in economic and financial matters than the AS managers.

Table 5.4 Comparison of Costs of Agricultural Societies and Commercial Companies

	Yield Necessary to Cover Costs (kg/ha)	ASs in Comparison with CCPCs (%)	Output Price Necessary to Cover Costs (lei/kg)	ASs in Comparison with CCPCs (%)	Managers Estimation of Production Costs (lei/kg)	ASs in Comparison with CCPCs (%)
Wheat						
All farms	2,796		2,507		1,847	
ASs	2,953	12.5	2,666	15.5	2,032	31.6*
CCPCs	2,625		2,308		1,544	
Sunflower						
All farms	1,662		3,193		2,550	
ASs	1,643	−2	3,181	2	2,800	28.6*
CCPCs	1,673		3,238		2,177	

* Statistically significant at 5% level

Another explanation concerns the difference in management between the two types of farm structure. The two types correspond to two different decision-making patterns. Whereas this is presumably hierarchical for CCPCs, it may be either hierarchical or democratic for ASs. In fact, the mode of governance varies from one society to another. The observations made during the field research tend to indicate that a democratic mode of governance was more often the case. The AS may be governed as labor-managed firms. Zusman (1993) points out that this may be the cause for some inefficiency. He underlines the fact that membership heterogeneity, which is clearly the case in the ASs in Romania, "creates considerable temptations to misuse the collective arrangements" and points out that "elected officials may derive personal advantage at the expense of other members" (1993, 47). Among the sixteen ASs interviewed, the number of members varied from 111 to 1,100, with an average of 429 members. So the number of members was always quite high, especially when compared to the number of permanent employees, which varied from two to forty-five. Each member owned a small land area and had, therefore, little influence on the AS as a whole, and the control on management was very fragmented. Free riding may occur in such cases and incentives to the manager to improve efficiency are very weak.

If the higher production costs persist, and unless transaction costs are lower in AS (a fact that tended to be contradicted by anecdotal evidence), this might undermine the viability of associations. Christiaensen and Swinnen (1994) emphasize the need for minimizing production and transaction costs, and that structures that fail to satisfy this condition will be outcompeted by those that do.

Farmer–Banker Relationship

The main problem that appeared to impede farmers' access to bank credit, thus the main external constraint, was that the minimum rate at which bankers agreed to lend to commercial farmers exceeded the maximum rate at which farmers agreed to borrow. Most banks were willing to lend at between 45 and 65 percent, while the farmers' demand for credit was strongly reduced when interest rates charged were above 40 percent. Thus, the interest rate charged by banks was shown to be above the maximum interest rate that farmers were willing to pay.

This, however, does not automatically mean that banks were not willing to lend to agriculture, since the banks might not have been fully responsible for the rate of interest that they charged to their customers. They might have suffered from financial repression and become price takers, rather than price makers. Interviews conducted with the branches of banks in both Ialomiţa and Teleorman, however, confirmed that banks were lending little or nothing at all to agriculture. Whereas the situation had been fairly stable in 1998 and 1999, the volume of lending was strongly reduced in 2000, both in terms of the number of farmers who had access to bank credit for variable inputs (seven in 2000 as opposed to seventeen in 1999) and in terms of the amounts borrowed (lei 162 million per borrower in deflated terms in 2000 as compared to 237 million in 1999). Banca Agricolă, the main bank, was being reorganized and in 2000 ceased making loans for a few months. BIR, which entered the market in recent years, ran into problems in February 2000. BRD declared that it wanted to stop making loans to agriculture and start lending to agri-food businesses. To counterbalance all these departures from the market, only one bank was reported to enter this market, CEC in Slobozia.

Are Bankers Willing to Lend?

Table 5.5 illustrates how the amount of total loans outstanding from the Banca Agricolă has decreased in recent years. Commercial farms, however, may be even worse off than the general lending situation shows. BRD, for example, repeated on several occasions that it did not wish to lend to farms anymore and BCR appeared to be focusing on agri-businesses, rather than on farm companies. Banca Agricolă was very restricted in its activities in 2000, due to its reorganization, but table 5.5 shows how its supply of loans had decreased since 1996. For some banks the situation may evolve, so the availability of credit to commercial farmers may develop in coming years, but for the time being the situation is highly unfavorable to commercial farmers.

Table 5.5 Banca Agricolă's Evolution of Outstanding Credit, 1996–1999 (adjusted for inflation: 1996 = 100)

(Million lei)	1996	1997	1998	1999
Total assets[1]	15,025,995	4,682,324	3,431,349	1,493,325
Loans and advances to customers[2]	8,926,376	1,397,049	971,126	187,475

[1] Bad loans were taken out of BA portfolio in April, 1999 and transferred to a special purpose vehicle

[2] Including assets being sold off

Source: Banca Agricolă *Annual Reports* (1996–1999)

At first glance, there were inconsistencies in the answers from the banks. On the one hand, they reported zero or only a few bad loans, but, on the other hand, they showed an apparent lack of willingness to lend to commercial farmers. Two issues emerged from the interviews that may provide some explanation of the banks' position. First, the concept of "objective" as opposed to "subjective" loan default seemed to be fairly well shared among bankers in Romania. By "objective" loan default, bankers mean a willingness but impossibility to repay (usually due to crop failure), while "subjective" loan default means an unwillingness to repay. This is important, as the bankers may act as though they perceive a high risk of "objective" loan default although the "subjective" rate is low.

The other problem concerns collateral. The types of collateral acceptable to farmers and bankers, respectively, were analyzed in detail. There appeared to be a lack of match between the two in several cases. Banks often would not accept land or future crop as collateral, whereas farmers would like to pledge them. On the other hand, banks would favor a personal guarantee from the farmer or his or her family, or real estate in town, which the farmer often either did not want to or could not provide. A continuing relationship between farmer and banker was assumed to ease this situation. Three indicators were used to analyze this relationship: the existence and length of a borrowing relationship; the number of visits per year paid by the farmer to the bank; and the number of visits per year by the banker to the farm. Whereas the existence of a good credit record (i.e., a borrowing relationship of five years or more, and borrowing in the previous year) appeared to make a significant difference to the provision of a loan, this was not the case for the frequency of visits by either the farmer to the bank or the banker to the farm. Frequent visits could just be an indicator of debt crisis management.

In addition, some bankers mentioned that guarantees were at best difficult, if not impossible, to enforce. In the context of asymmetric information, this may point to an issue of moral hazard and would explain why only regular customers received loans. A well-organized system of warehouse receipts was cited as one means of alleviating this problem.

Are Romanian Banks Repressed?

As mentioned above, the National Bank of Romania (NBR) increased to 30 percent the rate of reserve that it required on deposits after December 1999. In parallel, it promised to raise its compensatory interest rate from 16 percent to 30.5 percent. In summer 2000 the rate of interest quoted by the banks for term deposits of up to six months varied from one bank to another. If the National Bank pays interest on reserves, the minimum rate at which banks must lend to break even is between 44 and 51 percent. However, if the National Bank does not pay interest on the mandatory reserves, the break-even rate increases into the range of 57 to 64 percent.[1] The results under different alternatives are shown in table 5.6.

Table 5.6 Lending Rates Required Given Different Rates Paid on Deposits and Compensatory Interest Rates (percentages)*

		Compensatory Interest Rates Paid by the National Bank		
		0	16	30.5
Interest paid on	40	**57**	**50**	**44**
deposits	45	**64**	**57**	**51**

* Lending rates required are shown in bold.

To the extent that banks are free to charge the interest rate that is necessary to break even given the conditions imposed by the NBR, they cannot be qualified as "repressed." However, since the study found that there was no demand for loans at rates above 55 percent, it could be that the imposition of mandatory reserves created a gap between what borrowers were willing to pay for loans and what the banks needed to charge. This created impediments to credit for agriculture. The situation was aggravated by the lack of unconditional enforcement of compensatory interest paid on the reserves. For example, for June 2000 the NBR changed the interest rate on minimum reserves and set them at 26 percent for the reserves in lei, but one bank reported that only 16 percent had been paid on these reserves.

Are ASs Rationed?

The lending market appeared to be quite small for everybody and access to credit difficult for all farmers. Yet, the study tends to show that it is even smaller for ASs than for CCPCs. Comparing the two farm structures, ASs were either more reluctant to borrow or gave up requesting a loan more often than CCPCs, while no difference could be found in the apparent credit needs, that is, in the amount of additional variable inputs that they wished to use. One explanation is the mentioned differences in the decision-making process between the two farming structures.

In addition, when a request was made, ASs were denied access to bank credit (table 5.7). Although eight ASs formally requested a loan, only two of them had some debt outstanding in 2000. One had not obtained this credit in 2000, but had unpaid arrears from a loan made in the previous year. Three ASs who had borrowed in 1999 and could reportedly produce adequate collateral were not granted a loan in year 2000. As a result, only one ASs got a loan in 2000, and this was only for part of the amount it had requested.

Table 5.7 Loan Outcome by Legal Status and Bank's Answer to Loan Request

Legal Status	Number of Loan Requests	Answer from the Bank		
		Loan granted in full	Loan partly granted	Loan refused
ASs	8	0	1	7
CCPCs	8	5	0	3

In Teleorman, for example, Banca Agricolă indicated that a main reason for rejecting four out of six credit requests was that it was severely credit constrained. These four loans were all requested by ASs. At the same time, the same bank granted credit to three of the CCPCs in the sample.

Farmers' Access to Suppliers' Credit

In total, twenty-six farmers requested credit from their suppliers. One farmer was refused, seven got the total amount that they asked for, and eighteen received part of their request. Therefore, it appears that the suppliers' credit eased the credit constraint on farmers. However, the terms and conditions applying to suppliers' credit were not transparent. Only twenty-one farmers could say something about the costs of suppliers' credit, but their answers were not always convincing, given the information collected in the field. From the suppliers' price lists that were gathered during the survey it appeared that their general conditions for credit consisted of a 5 percent fee up front (i.e., the prices of inputs were increased by 5 percent unless the customer paid cash). The foreign exchange risk had to be borne by the customer (all but one of the farmers indicated that the suppliers' credit was denominated in U.S. dollars). Farmers were supposed to pay off their debt as soon as they harvested the crop. Taking into account the 32 percent devaluation of the lei against the U.S. dollar in 1999–2000, from 16,700 lei/U.S. dollar in October 1999 to 22,000 lei/U.S. dollar in July 2000, and the 5 percent charge up front, this 37 percent cost over the ten-month period amounted to an annual rate of about 44 percent. Since the farmers, however, preferred to calculate in terms of actual disbursements rather than on an annual rate basis, they should have quoted a cost well in excess of 30 percent. Only six farmers produced a realistic estimate, framing their suppliers' credit cost between 25 and 40 percent.

How well does suppliers' credit perform as a substitute to a bank credit? According to the New Institutional Economics, one of the main competitive advantages of suppliers over bankers in a situation of asymmetric information is the amount of information that they may be able to gather. The flow of information is presumably better between farmers and suppliers than between farmers and bankers, and suppliers can better identify the "good" customers from the "bad" ones. Furthermore, suppliers may more readily "retaliate" against bad payers than do bankers, simply by stopping further supply in cases of non- or partial repayment, while bankers need a court judgment to enforce the law.

While these two advantages are important, suppliers also bear substantial transaction costs. Usually, one advantage expected from interlinking markets is lower transaction costs. However, in the case of Romania the transaction costs incurred by suppliers are still quite high. As they import most of the inputs that they sell, they are obliged by the circumstances of the unstable macroeconomy in Romania to invoice their customers in a foreign currency, most often in U.S. dollars, whereas their customers' only source of revenues is in Romanian lei. They draw no benefit from the exchange rate fluctuations that, however, increase the risk incurred on these credits. The bigger the fluctuation, the higher the repayment cost for the customer, an additional expense that may or may not be compensated for by an increase in the price of the crop sold. This also increases the risk for input suppliers. To monitor their credit, suppliers therefore are often obliged to be present in the fields at the time of the harvest, which increases the monitoring costs.

To a certain extent, suppliers' credit in Romania eases the constraints on credit for farmers. It certainly allows transactions to take place where they might have failed due to the lack of credit. But interlinked credit also has some drawbacks, the main one of which may be the lack of competition. The impressions from the field survey were that the dominant presence of a few suppliers might maintain a market with little or no competition and high prices. For example, although farmers routinely mentioned that they had between three and five suppliers, they may have only one per type of input.

In summary, it appears that while suppliers help with transactions in a market that would have otherwise failed, this involves substituting for banks and dealing with financial problems, such as managing foreign exchange positions for which suppliers are less well prepared than the banks. In addition, it does not appear that suppliers participate in creating a transparent and competitive market.

Conclusion

The post-transition environment in Romania has not been favorable to the development of bank lending to agriculture, either at the macroeconomic level, because of the unfavorable levels and relations between interest rates, or at the microeconomic level, since farm structures are still in a process of transition and borrowing practices are not rigorous.

Farmers' credit in Romania is limited by a combination of internal and external constraints. Out of twenty-nine farmers studied, thirteen farmers did not get a loan because they did not ask for one, and thirteen were refused the loan they requested. Only three farmers got a loan. As a result, bank credit had low importance for the provision of funds for variable inputs, while suppliers' credit appeared to be of crucial importance. However, suppliers only provided the bare minimum inputs necessary to crop the fields. Disinvestments continue, which can undermine the future of the commercial farms.

The short history of commercial farmers in Romania acts as an impediment to their access to bank credit. Most Romanian banks do not wish to lend to agriculture at the present time. Many banks have disappeared from the lending market altogether, because of financial problems. Farmers are credit rationed. Farm structure apparently matters, since banks were found to refuse a loan less often when one was requested by a CCPC than when requested by an AS.

Several policy implications follow from this study. They are not narrowly focused on credit. Rather, it is believed that the thrust of the policymakers' efforts must focus on upgrading the financial paradigm in which both farmers and bankers operate. To become better (i.e., frequent and reliable borrowers), farmers need to be better informed. They must be able to assess the financial consequences attached to price variations of the inputs they use and to different production yields or market prices. For this purpose, it is important to expand extension service advice to incorporate farm accounts and budgeting and help farmers make forecasts under different scenarios of yields and prices. Dissemination of price information also needs improvement. In addition, it is necessary to identify alternatives to traditional collateral, for example, certified warehouse receipts, and to work out the detailed legal framework and technicalities that would make them really enforceable.

Notes

1. When a reserve requirement k is imposed on a bank for its deposits D, a volume $k \times D$ of the reserve is kept in the accounts of the National Bank, while $(1 - k) \times D$ could be lent in the market. It is assumed that the bank pays a nominal interest rate i_d on its deposit base D and charges a nominal rate i_l on its loans L. To break even on its loan portfolio, the bank should receive as much as it pays, that is:

$$i_l \times (1 - k)D = i_d \times D$$

If the National Bank pays a compensatory interest rate i_c to the commercial bank on the volume of deposit that it holds in mandatory reserves, then the equality becomes:

$$i_d \times D = [i_l \times (1 - k)D] + (i_c \times kD)$$

That is, solving for i_l:

$$i_l = i_d - (1_c \times k)/(1 - k)$$

References

Banca Agricolă. 1996. *Annual Report 1996*. Bucharest, Romania.
———. 1997. *Annual Report 1997*. Bucharest, Romania.
———. 1998. *Annual Report 1998*. Bucharest, Romania.
———. 1999. *Annual Report 1999*. Bucharest, Romania.
Christiaensen, L., and Swinnen, J. 1994. *Economic, Institutional and Political Determinants of Agricultural Production Structures in Western Europe*. Working paper 1994/11. Leuven, Belgium: Department of Agricultural Economics, K. U.
Gavrilescu, D., and Sarris, A. H. 1996. Restructurarea fermelor li sistemelor agricole din România. In *Restructurarea Agriculturii li Tranzitia Rural în România*, ed. D. Gavrilescu. Bucharest: Agris Redactia revistelor agricole.
Heidhues, F. and Schrieder, G. 1997. Access Constraints of Romanian Peasants in Relation to the Formal Financial Sector. *CERT Discussion paper 97/12*.
McKinnon, R. I. 1991. *The Order of Economic Liberalization: Financial Control in the Transition to a Market Economy*. Baltimore and London: The Johns Hopkins University Press.

Ministry of Agriculture and Food, World Bank and the European Commission. 1997. *Private Agriculture in Romania, Farm Survey*. Bucharest, Romania. Mimeo-graphed.

National Bank of Romania. 1999. *Annual Report 1999*. Bucharest, Romania.

———. 2000. *Interim Report June 2000*. Bucharest, Romania.

World Bank. 1996. *Romania Farm Survey*. Unpublished Report on a Survey of Private Romanian Farms. Bucharest, Romania.

Zusman, P. 1993. Participants' Ethical Attitudes and Organizational structure and Performance: Application to the Cooperative Enterprise. In *Agricultural Cooperatives in Transition* eds. C. Csaki and Y. Kislev., Boulder, Colorado: Westview Press.

Chapter 6

A Model of the Impacts of Agricultural Policies on Farm Incomes

Nigel Williams, Paul Hill, and Adriana Cristoiu

There are many different policy measures that can be deployed in the quest to improve the income of farmers. Such policies vary in their cost and in their impact on different parts of the agricultural sector, according to size and type of farm and according to prevailing economic conditions. In order to make the most effective use of the limited funds available for supporting agriculture, it is important to have tools that can be used to assess the impact of any proposed policy measure on farm incomes.

The aim of this project was to provide a tool for studying the effects of a range of different policy options on farm incomes. In order to do this, a methodology was defined, a model was designed and tested, the necessary data were collected, and then different policy options were examined using the model. The results from the model provided valuable insights into the effectiveness of the various policy options examined and also on their impact on the different sectors of Romanian agriculture.

Methodology

Within the EU, data on farm incomes are routinely gathered through the Farm Accountancy Data Network (FADN)[1]. In England, for instance, FADN collects 2,350 farm records per year, a sampling fraction of nearly 2 percent from a population of 130,000 full-time farms. The approximate cost of the survey, including management, data collection, analysis, and support services is U.K. pounds 2.25 million per year or in the region of U.K. pounds 1,000 per farm. As Romanian agriculture comprises about four million farms, to achieve even a one percent sampling fraction would require collection of data from 40,000 farms. The cost of such a survey would be prohibitive.

The shortage of resources, the large number of holdings, and the absence of detailed and up-to-date information on the different sizes and types of farm in Romania precluded an EU-style FADN methodology. It was important to develop an alternative model that would enable the researchers and policymakers to gain insights into the potential effects of different policy options quickly and at low cost. It was also important that the model should be capable of development and refinement through time as resources permit. Given the above considerations, it was decided to survey a small number of farms and to use them as in-depth case studies (Barnard 1975). While this approach is not ideal, it is quick, low cost, and does provide information on how proposed policy measures will affect real farms.

The methodology employed was to use spreadsheets to build relatively simple financial models of a selection of representative farm types. The financial models were based on standard accounting conventions for agriculture, thus they included profit and loss statements, balance sheets, and quarterly cash flows; all elements were fully reconciled to each other to ensure accuracy and internal consistency. These models were then used to investigate the effects of a range of possible policy options on the different farm types. It is important to note that the models are static. The underlying assumption is that the policy measures being tested do not influence the mix of enterprises or the levels of input use and technical efficiency of the farms. This is a realistic assumption in the short term, not least because of the fixed nature of many agricultural resources. The long production cycle and the experience of the farmers tend to preclude rapid and/or substantial changes in farming systems in response to external stimuli in the short run. Longer run effects of policy measures can be investigated through the use of optimizing models, but these are more expensive and time consuming to construct and require more data.

Data Collection

In order to investigate the effect of proposed policy measures on as wide a range of farm business types as possible, a group of ten case study farms was selected. The farms were all located in southern Romania to minimize travel costs. The group consisted of two CCSCs, two ASs, and six individual farms. The six individual farms were further subdivided into pairs on the basis of their main source of income, that is, agriculture, salaries (paid employment), and pensions. The logic behind the selection was that CCSCs and ASs were important, despite their relatively small number, because of their large size and the high proportion of total production deriving from them. Individual farms were important because of their number, and the subdivision within this group reflected the diversity of income sources available.

Each of the selected farms was visited by a member of the team during the spring of 1998. The data collected incorporated details of the agricultural production systems, including yields, sales, and home consumption of produce; expenditure on inputs, including capital items, household expenditure, and levels of indebtedness; and estimates of the value of capital assets, including land, buildings, and machinery. As far as was possible, valuation procedures were consistent with management accounting principles and the FADN rules: all livestock and agricultural produce were valued at conservative market values, stores of purchased inputs were valued at cost, and land was valued using information from the informal land market. The only divergence from these principles was in the valuation of buildings and machinery that, in the absence of better information, were valued on an historic cost rather than on a replacement cost basis.

The Farm Model

The model was based around a harvest year, that is, a crop production cycle, rather than an accounting year. The aim was, as far as possible, to incorporate the crop production costs in the same financial year as the output, thus minimizing problems of valuing unsold produce at the year end or growing crops at the start of the year. For administrative convenience, the year end on the CCSCs and

ASs was taken as the end of December, and for individual farms it was the end of October. Any stocks of produce remaining unsold or unconsumed at the end of the chosen twelve-month period were entered into the closing valuation at net realizable value.

The first stage in the modeling was the preparation of gross margins for each individual livestock and crop enterprise. Livestock enterprise output included sales of produce and was adjusted for breeding herd/flock depreciation. Crop enterprise output was calculated by adding the sales of that year's crop to the value of unsold crops in store at the year end. Internal transfers of salable produce (e.g., cereals consumed by own livestock) were valued and added to both output and variable inputs. Produce consumed in the home was added to output and also to private drawings. This gave a more accurate indication of the level of farming output on the holdings, especially for the individual farms where a significant proportion of the production was for home consumption. Thus, for each enterprise *i*, output was calculated as follows:

$$O_i = P_i (Y_i + H_i + T_i - R_i) + CV_i - OV_i$$

where

O_i = output of enterprise *i*

P_i = the price per unit of output *i*

Y_i = quantity of produce sold

H_i = quantity of produce consumed in the home

T_i = quantity of produce/number of animals transferred for use in another enterprise (e.g., seed, feed grain, or calves for rearing)

R_i = number of animals transferred in from another enterprise (e.g., heifers in calf transferred into a breeding/production enterprise from a rearing enterprise)

CV_i = value of stock at the end of the year

OV_i = value of stock at the start of the year

Variable costs (feed, veterinary and medicine costs, seed, fertilizer, pesticides, casual labor, and machinery services) were then allocated to each enterprise and deducted from output to give enterprise gross margin. These individual gross margins were summed to give total farm gross margin. Fixed costs (regular labor, own machinery costs, farm maintenance costs, rent, and miscellaneous overheads) were then deducted from the total gross margin to arrive at an estimated "profit" or trading surplus for the year. Thus, profit was calculated as follows:

$$\Pi_i = \Sigma(O_i - VC_i) - FC$$

where summation is over all enterprises *i*,

Π_i = trading surplus for the year
O_i = output of enterprise *i*
VC_i = variable costs of enterprise *i*
FC = total fixed costs

Any interest payments were deducted from the trading surplus to show a surplus after interest. Balance sheets for the start and end of the financial year were drawn up using data on assets such as land, farm buildings, machinery, livestock, stocks of harvested crops, and purchased stores. Financial assets, such as cash at bank and debtors, and liabilities, such as long, medium, and short term loans, trade creditors, and bank overdrafts, were also recorded. Because of valuation problems, the value of the family house was excluded from the balance sheet on household farms.

The data collected were then used to draw up quarterly cash flows for the twelve-month period under study. It is important to emphasize that the cash flow recorded the timing of cash transactions rather than the physical movement of outputs and inputs. Thus, produce which was moved off the farm in the final quarter of the period but for which payment was not received until after the end of the period would not appear in the cash flow; instead it would appear in closing debtors. Equally, although the gross margin calculations applied to the harvest year covered, disposal of opening stocks of produce within the period would appear in the cash flow, as would any payments of opening debtors and creditors. The cash flow statements were important as they permitted calculation of the level of indebtedness or otherwise of the farms on a quarterly basis. Interest receipts and charges were calculated on positive and negative cash balances, respectively, and added to or deducted from the farm's trading surplus.

Internal consistency between the three elements of the financial model was ensured by reconciliation of the profit and loss account to the balance sheet and of the net cash flow to the profit and loss account.

Finally the results were normalized to remove any unusual or atypical variation in outputs or costs of a farm. An example would be where a freak hailstorm had damaged a top fruit crop, rendering it virtually unsalable. In such a case, the expected yield was substituted for the actual yield and output, gross margin and trading surplus adjusted accordingly. However, where the yield recorded fell within the normal range of variation to be expected, for example, a top fruit crop grown in an area subject to regular frost, the gross margin was not adjusted.

Results

The Initial Results for Individual Farms

The initial results of the survey of the farms provided useful information on their financial condition. It also provided indications of the relative importance of sources of income from outside agriculture for the individual farms and the extent to which agriculture contributes to funding farm family household expenditure. As can be seen from table 6.1, the individual farms were small in area. The smallest were the two pensioner holdings. Not only were they small in area terms, but their production from agriculture was low. The farm profit was insufficient to meet the household needs, in one case being only 18 percent of household consumption. The farmers with salaries operated larger farms, and profit from farming accounted for something over half of household expenditure. One of the two farms without paid income earned a significant proportion of income from a village shop. The other obtained all its income from farming. In several cases the produce consumed in the home represented a significant proportion, and in two cases, over half of total household consumption. Four of the farms had a level of private income that exceeded their farm profit, but even after adding other income to farm profit, two of the farms did not generate sufficient funds to cover their household expenditure.

In all cases the farms are debt free and so have a percentage equity of 100 percent. As a result, there are no interest expenses to deduct from trading surplus. The profit expressed as a percent of net capital ranged from 5 percent to 23 percent. This apparently healthy return should be considered in the context of an inflation rate of over 150 percent prevailing during 1997. Therefore, the real rate of return (adjusted for inflation) was negative in all cases.

Table 6.1 Basic Data for the Six Individual Farms, 1997 Harvest Year ('000 lei)

Farm Type	Pension 1	Pension 2	Salary 1	Salary 2	Farm 1	Farm 2
Area (ha)	1.8	1.07	4	6	5	3
Gross margin	2,530	1,649	22,453	9,175	9,803	4,548
Trading surplus from farming (profit)	1,630	1,109	23,686	8,974	9,803	3,862
Private income	2,079	5,651	19,506	10,800	104	17,529
Total income	3,709	6,760	43,192	19,774	9,907	21,391
Household expenditure	1,560	4,262	34,400	14,000	4,000	5,200
Produce consumed in home	2,700	1,982	10,183	972	6,054	1,659
Total consumption	4,260	6,244	44,583	14,972	10,054	6,859
Income surplus or (deficit)	(551)	516	(1,391)	4,802	(147)	14,532
Profit as percentage of consumption	38%	18%	53%	60%	98%	56%
Net capital	34,930	11,400	149,828	38,761	55,719	62,985
Farm profit as percentage of capital	5%	10%	16%	23%	18%	6%

The Initial Results for Commercial Farms

In contrast to the individual farms, the ASs and CCSCs were largescale and, at least potentially, commercial operations. The data for these farms have been converted to a per hectare basis to preserve confidentiality.

Table 6.2 Basic Data for the Four Large-Scale Farms, 1997 Harvest Year ('000 lei per ha)

Farm Type	AS 1	AS 2	CCSC 1	CCSC 2
Gross margin	2,683	1,020	3,263	4,098
Fixed costs	665	291	1,152	2,944
Trading surplus (profit)	2,018	729	2,111	1,154
Other income	0	0	142	706
Interest paid/(received)	(356)	(5)	375	502
Profit after interest	2,374	733	1,878	1,358
Liabilities	0	351	3,288	1,423
Net capital	7,862	602	2,653	5,051
Total capital	7,862	954	5,941	6,473
Profit as percentage of net capital	30%	122%	71%	27%
Tax payments	593	183	660	516
Distributed profits	598	702	59	165
Total withdrawals	1,191	885	719	681
Withdrawal as percentage of profit	50%	121%	38%	50%

The large-scale farms were characterized by more intensive production than on the individual farms; they were also more likely to borrow money for farming purposes. Income from nonfarm sources was much less important in percentage terms than on the individual farms and the return on net capital was generally higher. After-tax profits could be reinvested in the business and/or distributed to members in ASs. In the case of one AS, the distribution of profits to members exceeded the total farm profit with a resultant fall in net capital in the balance sheet. This is unsustainable if carried on for any length of time and reflects the problems sometimes associated with cooperative-type ventures, where the short-term personal interests of the individual members who wish to maximize their disposable income can be in conflict with the longer-term needs of the business.

Updating Results to the 1998 Harvest Year

As explained, the financial results obtained from the case study farms were for the 1997 harvest year. The timing of the study during the summer and autumn of 1998 precluded the collection of data for that harvest year, and delaying the study until such data were available would have reduced the usefulness of the policy analysis that was to be carried out. A feature of the two-year period covered by the 1997 and 1998 harvest years was the cost/price squeeze caused by the high levels of inflation and the differential rate of price changes for outputs and inputs. Failure to update the 1997 figures for these changes would have rendered the analysis of the policy instruments much less meaningful or would have even made them misleading. In order to provide a more up-to-date analysis of the possible policy measures, the results from 1997 were reworked to provide estimates of performance in the 1998 harvest year. Quarterly price index numbers were applied to inflate the 1997 cash flow data to 1998 values. This procedure gave an updated version of the cash surplus/deficit and also of any interest payments. The balance sheet was also indexed up for inflation using the appropriate index numbers. Since the cash flows and balance sheets were fully reconciled to the trading account, it was then possible to combine the results from these two updated financial statements to calculate the trading surplus that would have been earned under the input and output prices prevailing in 1998, using the 1997 farming system.

Policy Options

Once the models had been updated to incorporate the expected price changes, various policy instruments were processed through the models, and their effects on a number of key indicators were examined. The policy instruments considered for testing were:

- vouchers (input subsidies)
- credit subsidies (interest rate subsidies)
- output price support

The key indicators identified for analysis included change in net cash flow, change in trading surplus before interest, change in trading surplus after interest, and change in net capital. In the following description, attention is focused on the two measures of trading surplus, as there is considerable interaction between these measures in a static model. Any improvement in trading and total net cash flow will either reduce interest payments or increase interest earned. Any increase in the trading surpluses will automatically increase net capital by the same amount.

Vouchers

The voucher scheme (introduced in chapter 2) was designed by the World Bank as a means of providing input subsidies in a non-distorting way. Due to the poor records regarding who was actually farming what land, the vouchers were allocated to private landowners rather than to farm operators. However, the vouchers were tradable so they could be passed on by the owner to whoever was cultivating his or her land, irrespective of whether they were individuals or associations. Within the model, the value of a voucher was assumed to be lei 230,000. The payment was split between autumn and spring to match the main periods of expenditure (on sowing and on harvesting). The following assumptions were also made:

- Individual farms were assumed to receive the full value of the vouchers appropriate to the size of holdings. While some small farms might not spend sufficiently on inputs to be able to use vouchers directly, it was decided that, as the vouchers were fungible, the households would be able to obtain the value of the vouchers by trading them.
- ASs—all members of the society were assumed to be eligible for the vouchers on their land and to pass their vouchers to the society.
- CCSCs—these farms were not eligible for vouchers. However, this option was tested for CCSCs as well, in order to compare the effect of different policy options across all the farms. It was assumed that these farms would receive one voucher for each hectare.

The effect of this policy instrument was an increase in the value of closing balances for all the farms. This effect was predictable. It has been shown (Lazaroiu, Luca, and Tesliuc 1998) that providing these vouchers was a way of supporting the incomes of the private farmers. The effect of this policy was most marked on the individual farms. Compared to the initial no-policy position, profit increased on all farms. The increase ranged from about 3 percent in the case of a salary farm to over 20 percent for a pension household. Although there was some improvement in income on the ASs and CCSCs, this was not as great as that experienced with the other policy measures tested. It should also be remembered that these farms were assumed to receive one voucher for each hectare farmed, which is almost certainly overstating the extent to which they (especially the CCSCs) would be eligible.

Interest Subsidies

The second policy instrument tested was a subsidy of 70 percent on the interest of short-term loans. This policy instrument was chosen for testing because a law regulating this was adopted by the Romanian Parliament in the summer of 1998, to be effective for the following three years. The funds made available could be used to finance expenditure, including animal breeding activities. The subsidies were to be available to agricultural holdings irrespective of their status, thus applied to individual farms, ASs, and CCSCs.

The results show interest subsidies had the greatest impact on the legal entities (CCSCs and ASs). This is unsurprising because these farms operate as large-scale commercial organizations and so it is common practice for them to borrow money from banks. As would be expected, the farms that reaped the greatest reward for this policy option were those that borrowed the most money. These were the large-scale commercial farming operations that typically borrowed money as short-term loans to fund the working capital (seed, fertilizers, and pesticides) needed for modern farming operations. The largest change in profit due to this policy option was an increase of 62 percent. The ASs under study were not affected by this policy measure because the two case study farms did not borrow. This is not always the case with ASs but, as emphasized in chapter 5, societies are often either reluctant to borrow or denied access to credit. The policy had little impact on the individual farms. Again this was to be expected, since such individual farmers prefer to borrow money from informal sources (relatives or friends), cooperative credit institutions, or suppliers.

Output Price Support

Output price support can take many forms: it can include deficiency payments, support buying, export refunds, and intervention storage, as well as setting a minimum price by law and giving subsidies to processors.

For the purpose of this model, a one percent subsidy on all agricultural production, including produce unsold at the year end and internal transfers to livestock, was implemented. This policy measure was applied to all farm types. In all cases, the trading and total net cash flows increased. For some individual farms the improvement from the initial situation was small when compared to the effects of vouchers. This was because the contribution of trading to the total income in these cases was low, as they were largely dependent on off-farm sources of income. For CCSCs and ASs the effects of price support were also positive, as would be expected.

Discussion of the Results

The models produced a great deal of information on the way in which the different policy options affected the case farms. A verbal summary of the effects of the policy options as measured by a selection of financial indicators is set out in table 6.3, followed by a numerical summary in tables 6.4 and 6.5.

Table 6.3 Effects of the Policy Options on Financial Indicators

Indicator	Policy Instrument		
	Vouchers	Interest Subsidy	Output Price Support
Net cash flow—quarter one	Improvement in net cash flow equal to the value of the voucher with associated reduction in borrowing and interest payments.	Some effect on net cash flow on indebted farms, none on farms with positive credit balance.	Nil effect on quarterly net cash flow.
Net cash flow—quarters two and three	Reduction in borrowing and interest payments.	Some effect on net cash flow on indebted farms, none on farms with positive credit balance.	Nil effect on quarterly net cash flow.
Net cash flow—quarter four	Improvement in net cash flow equal to the value of the voucher with associated reduction in borrowing and interest payments.	Some effect on net cash flow on indebted farms, none on farms with positive credit balance.	Improvement in net cash flow, depending on timing of sale of produce. Little effect on farms where majority of produce is consumed in the household.
Trading surplus	Increase equal to value of vouchers.	Nil effect.	Increase equal to value of output subsidy.
Trading surplus after interest	Increase in trading surplus plus reduced interest payments throughout year.	Value of interest subsidy plus associated reduction in interest paid on reduced borrowings.	Increase in trading surplus plus reduced interest payments in last quarter.
Private drawings	Nil effect.	Nil effect.	Increase in opportunity cost of home-consumed produce caused by price support.
Net capital	Increase equal to trading surplus after interest less private drawings.	Increase equal to trading surplus after interest less private drawings.	Increase equal to trading surplus after interest less private drawings.

Table 6.4 Percentage Change in Trading Surplus on the Individual Farms as a Result of the Policy Measures

Farm Type	Pension 1	Pension 2	Salary 1	Salary 2	Farm 1	Farm 2
Vouchers	11.3	17.3	2.5	12.3	9.3	11.0
Interest subsidy	0.0	0.0	0.0	3.0	0.0	0.0
Price support	1.7	0.0	0.6	2.0	0.9	1.2

Table 6.5 Percentage Change in Trading Surplus of the Large-Scale Farms as a Result of the Policy Measures

Farm Type	AS 1	AS 2	CCSC 1	CCSC 2
Vouchers	8.5	27.0	9.8*	22.6*
Interest subsidy	0.1	0.0	16.7	61.6
Price support	1.8	2.6	2.1	6.2

* Hypothetical result: vouchers are not available to CCSCs under current legislation as described above.

The use of vouchers clearly benefited the small individual farms. Even those which might appear to have little use for them because they purchased few inputs would be able to benefit from vouchers because of their fungibility. The impact of vouchers on the larger-scale farms was more variable, and it should be remembered that vouchers are not currently an option for CCSCs.

Credit subsidies will, by definition, only benefit those who borrow and who also enter into formal credit agreements. This largely excludes the small individual farms, which generally meet their finance needs through nonbank sources. One way to overcome this impasse for the small individual farm is to design a credit scheme that is more accessible to such households. When doing this, great care must be taken to balance the need to give farmers easier access to loans against the need to minimize default levels.

Price support benefits all, but the level of benefit varies directly with size of holding. Thus, small-scale farm households, especially those whose members consume a high proportion of their output in the home, show little benefit from this policy instrument; the extra value of output serves only to increase the opportunity cost of the produce consumed in the home. It is worth noting that, assuming there is no adverse effect on prices, a one percent improvement in technological efficiency has the same effect on farm income as a one percent increase in product price generated by a support measure. The added advantage of the technological change is that it should continue into the future at no additional cost, whereas the price support policy measure will require an equivalent level of public expenditure every year. The distributional effects of technological improvement are also more beneficial as, assuming that small farmers are able to adopt the technology, it will increase the trading surplus of the small-scale holdings and so increase disposable income rather than raising the monetary value of produce consumed within the household.

Conclusion

The results show clearly that different policy measures can have widely diverging effects on the case farms. This is because of the wide range of situations encountered on the farms. It is this richness of detail that makes the case study approach so useful for this type of analysis; detail is not lost in the process of aggregation to produce an "average" farm.

The results of the model confirm that policy measures should be tailored to the needs of the specific target groups—what is best for small-scale family farms is not appropriate for large-scale commercial holdings. This was most evident with the credit subsidies that had little impact on the income of the individual farms, but caused major improvements in the situation on the large-scale farms with regular borrowing requirements.

The fact that the results generated from the model indicate that some of the proposed policy measures will have little impact on the situation of householder farm families sends a clear message to policymakers. Alternatives to the traditional financial support measures should be examined in order to determine whether and to what extent they would meet the policy objectives of the government and also whether the policy objectives themselves need refining. Policymakers might wish to consider whether it is appropriate to have a single policy for two such diverse groups of farms as were examined here.

Romanian agriculture is characterized by a large number of small-scale holdings which cannot justify investment in machinery and sometimes even in variable inputs. In addition, the farmer may lack the knowledge to farm more effectively or may be unable to do so because of age or health problems. In such circumstances alternatives to direct income support may be appropriate, at least in the long run. Such alternative measures should be designed to encourage change to the underlying causes of the income problem. They may be institutional as well as financial; for example, reform of the land tenure system by means of a more formalized let-land sector to stimulate the consolidation of holdings.

The cost of implementing an FADN style survey in Romania is too great to be sustainable if it were to cover all farms. If it were to include only commercial farms, then most holdings would be ignored. Given the financial and other resource constraints facing the Romanian government, the methodology developed for this project offers a low cost and rapid means of gaining insights into the financial situation of both farm households and farm businesses. It also provides the opportunity for the effects of different policy measures to be investigated. The in-depth information available on the situation on the case study farms enables policymakers not only to see the effects of a proposed measure, but also to understand why the result has come about.

The model used in this project is capable of further refinement to include more data on household income and expenditure. The model is static, but this can be overcome if it is updated each year by the collection of the most recent financial data from the case study farms. The model does not fully implement current cost accounting principles, but could do so relatively easily, providing accurate data on replacement costs of buildings and machinery could be collected from the farms.

Notes

1. In French, RICA—Reseau d'Information Comptable Agricole.

References

Barnard, C. S. 1975. Data in agriculture: A review with special references to farm management research and advice in Britain. *Journal of Agricultural Economics* 26:(3) 289–334.

Lazaroiu, S., Luca, L., and Tesliuc, E. 1998. *Assessing the Impact of the 1997 Voucher Scheme.* World Bank Report. Mimeo-graphed.

Chapter 7

Price Stabilization Policies in the Romanian Wheat Market

Paul Webster, Lionel Hubbard, John Lingard, and Adriana Cristoiu

Price instability in the agricultural and food sectors is a major concern for governments around the world. However, confusion often arises between the objectives of "pure" price stabilization and farm price protection. In short, the former is concerned with stabilizing price around its long-run market equilibrium, whereas the latter is aimed at supporting price above this level and the associated trade-parity price. Economists tend to be skeptical about price stabilization policies and instead suggest hedging devices like futures markets and options. In other words, risk should be privately traded. In contrast, policy-makers often favor active market intervention that can distort the transmission of international price signals to the domestic market, particularly if the objectives of stabilization and protection become confused. As a compromise between these twin goals, Quiroz and Valdes (1997, 52) propose a floor price policy as a way of avoiding extremely low prices, and cite as one of their worked examples the Romanian wheat market. This proposal is taken as the starting point for an analysis of whether Romania should seriously consider introducing a price stabilization scheme for wheat.

The 2.4 million hectares planted to wheat represented 25 percent of the total arable area in Romania. The crop accounted for 8 percent of the total value of the country's agricultural exports, although in 1996 and 1995 the contribution was 36 and 22 percent, respectively, owing to both larger volumes exported and higher wheat prices. It is expected that wheat will continue to be a major part of agricultural output and will remain an important export as Romanian agriculture continues to adjust to a market economy.

The interaction of many factors produces variability of prices. Some of the important factors in the Romanian wheat market are:

- Macroeconomic and policy variables: world market conditions; exchange rate policy and the resulting dollar exchange rates; trade policies; and monetary policy, inflation rate, and interest rates.
- Market imperfections: the monopolistic structure of storage/drying/cleaning facilities; the lack of information about prices locally and at national level; the inability to achieve international quality standards; the lack of forward markets and inadequate financing mechanisms for stock holding; and the lack of competition among a monopsonistic set of buyers of wheat and suppliers of farm inputs.
- Domestic supplies of wheat: the growing conditions such as weather, diseases, drought; the availability of inputs and the finance to use them; and the profitability of alternative crops.
- Domestic demand for wheat: the change in demand for human consumption; population and income growth; the change in demand for animal feed; and livestock numbers.

All these factors affect current and future prices, and information relating to each will be involved in the formation of expectations about the level and variability of prices.

In the longer term, Romania must compete at world market prices without the need for intervention and will require a comparative advantage in wheat production in order to export. In the short run, even if price stabilization is more beneficial than hitherto generally acknowledged, it may not be an urgent priority for the improvement of farm level decision making in Romania, particularly given a shortage of appropriate skills and an untried and untested price stabilization system. Eliminating existing market imperfections with respect to, for example, finance, grain storage, and marketing, may give more immediate gains.

A brief analysis of the world market for wheat and its implications for Romania highlights the main trends and outlines some of the underlying reasons for the observed price patterns (for further details see Hubbard, Lingard and Webster, 2000). The world price has halved in real terms over the past forty years while its variability has increased. In the period 1960–1971 the world price of wheat, in real terms, fell at an average rate of 4 percent a year, and in the period 1976–1996, it continued to fall by almost 3 percent a year. The coefficient of variation (c.o.v.)[1] rose from 0.17 to 0.32 over the two periods, indicating increased price variability.

In part, these outcomes are a result of the agricultural policies pursued in many countries, and, in this respect, the EU must bear a major share of the blame. As for the future, commitments under the WTO will lead to reductions in the volume of, and spending on, subsidized exports and to more liberalized agricultural trade. Attempts to stabilize domestic wheat prices in Romania must therefore be seen in the wider international trade context of a strong and persistent downward trend in prices. Furthermore, international prices appear increasingly unstable, but proposed WTO agreements may modify the situation in the future.

The Impact of Price Instability on Romanian Farms

Price stabilization implies a reduction in the volatility or variance of prices. Price volatility is important in that it is associated with producer risk and therefore affects decisions. It is possible to distinguish two major effects. First, there is *anticipated* variability, which affects the way farmers plan their resource allocation. Examples are where risk-averse farmers might regularly underapply fertilizer to a crop whose returns are expected to be risky, or where investments are not made because of the impact of a poor outcome on repayment capacity. Second, there is *unanticipated* variability, which results in outcomes being outside the expected range and which immediately constrain subsequent decisions or may lead to hardship among those affected. For example, unexpectedly low prices may leave farmers without the ability to finance next year's crop or without the means to feed their families. Farmers make decisions on their own estimates of variability. These may be well or ill informed. The better informed farmers are about the risks that they face, the better their decision making will be, since losses due to unanticipated variability will be reduced.

Before discussing possible responses to price instability, it is helpful to examine the degree of variability faced by Romanian farmers in the recent past. Estimates of price variability for seven commodities based on monthly statistics between 1991 and 1995 are shown in table 7.1. The series covers a period of great structural change in the Romanian economy, but nevertheless provides some evidence of the prices which farmers have recently faced. The coefficient of variation for wheat is 0.26. This is higher than that reported by Quiroz and Valdes (1997) for world prices over the period 1986–1997 (0.17), but lower than the coefficient reported above for world price over the period 1976–1996 (0.32). Interestingly, all seven commodity groups show positive correlation coefficients for all combinations except for a small negative coefficient for pig prices with milk prices, implying that there was perhaps only limited scope for the reduction of risk through enterprise diversification during the period, an issue which is returned to. The crucial question is whether the anticipated degree of variability in the wheat price, which is not out of line with other objective evidence, has sufficient impact on decisions by Romanian farmers as to warrant the implementation of a stabilization scheme.

Table 7.1 Agricultural Price Variability in Romania

	Wheat	**Maize**	**Pota-toes**	**Live Cattle**	**Pigs**	**Broilers**	**Milk**
Mean	5.84	5.85	11.19	29.06	40.79	67.28	10.22
S. D.*	1.51	1.82	5.94	9.65	3.47	22.82	2.38
Variance	2.25	3.27	34.64	91.27	11.76	511.71	5.58
c.o.v.**	0.26	0.31	0.53	0.33	0.08	0.34	0.23
Correlation Coefficients							
Maize	0.98						
Potatoes	0.00	0.08					
Live Cattle	0.09	0.05	0.07				
Pigs	0.20	0.15	0.35	0.15			
Broilers	0.24	0.37	0.33	0.19	0.35		
Milk	0.18	0.37	0.20	0.18	−0.07	0.33	

* S. D. = Standard Deviation

** c.o.v. = Coefficient of variation

Source: World Bank, monthly prices (lei per kg) in private peasant markets, January 1991 to December 1995. Series deflated by Consumer Price Index, January 1991 = 100.

Responses to Price Variability

Experience has shown that farmers have many ways of coping with risk. The effectiveness of any strategy depends on the reduction in risk obtained and the cost of implementation. During June 1998, visits were made to a small number of farming enterprises and to grain market intermediary and storage companies in Romania. The aim in each case was to elicit existing risk management strategies, to gauge likely response to future instability in the wheat market, and to assess what survival strategies might be implemented. The following observations are based on these case studies. Further details of the individual businesses are given in table 7.2.

There are concerns among the managers about low prices and low levels of revenue to pay off debts, finance future crops, and support livelihoods, but variability, per se, appears to be of less concern since it implies periodic high prices as well. As for *diversification*, the case study farms have soils of excellent quality on which a variety of crops can be grown, and substitution would take place if the wheat price were low enough for a long enough period of time. A smallholder's response to instability and a lack of capital was not to grow wheat. As markets in general become more liberalized, it seems likely that the opportunities for alternative crops will increase. The potential to diversify against volatility in the wheat price seemed considerable for the case study farms, notwithstanding the positive correlation coefficients for prices reported in table 7.1.

Choosing *technology* that reduces risk is a strategy usually employed to reduce yield variability rather than price variability. However, since revenue (i.e., yield multiplied by price) variability is what tends to interest farmers most, one can expect to see such a strategy used where volatility is a problem. Theory suggests that risk-reducing inputs (e.g., water, pesticides) tend to be employed more, and risk-increasing inputs (e.g., nitrogen fertilizer) tend to be employed less. One of the case study farms would invest in its own storage, cleaning and drying facilities if funds allowed, and another has already invested in overhead irrigation equipment, which should have led to major reductions in yield variability (as well as increasing expected yields), but lack of working capital has, as yet, prevented its use. Shortages of *working capital* at realistic interest rates and over appropriate periods of time appear to overshadow the effects of price volatility in the allocation of variable resources.

Table 7.2 Characteristics of the Case Study Businesses

Type of Farm	Size in Hectares	Organization and Ownership	Cropping and Stocking	Disposal of 1997 Wheat Crop	Issues Affecting Timing of Disposal	Response to Issues Affecting Disposal
Commercial company with state capital (CCSC)	9,333	70% shares State owned, 22% by Investment Organization, 8% by private shareholders (364).	All arable. 31% wheat, 500 sheep.	13,000 t total. Half sold at harvest, rest sold in monthly lots up to February.	Early selling because of shortage of working capital. Could not use farm assets as collateral.	Reduction of proportion of wheat in the rotation.
Agricultural society (AS) 1	1,860	Ex-Agricultural Production Co-operative. 400 members.	All arable. 38% wheat, 9% members' plots.	3,000 t total. Two-thirds sold in October and November. Remainder to members.	Early selling because of pressure from the bank. Could only use ex-APC buildings and machines as collateral.	On such good soil (deep chernozem) it was likely that more profitable alternatives would be found.
Agricultural society (AS) 2	800	Ex-Agricultural Production Co-operative. 300 members.	All arable. 50% wheat.	1,140 t total. 600 t sold August to Romcereal. 200 t sold April to private miller. Rest to members.	Early selling at low prices to repay input loans. Late payment by Romcereal. Prompt payment by miller.	Reduction in inputs next year. Substitution of alternative crops.
Individual smallholder	3.5	Ex-steelworker who now had another full-time job doing shift work.	1.5 ha fruit. 1 ha maize, 1 ha sunflower.	No wheat. Maize fed to own livestock or sold to neighbors.	Would have liked to have sold maize to Romcereal, but quantity too small.	Looking for a collector's network to sell to. Wheat not grown because of high input costs.

A common means whereby total risk can be reduced is to spread it among many *shareholders*. This is the case for the Romanian CCSCs and ASs that have many members. Insofar as the members have other sources of income they can spread the risk associated with their shareholding. It could be argued that the search for foreign partners to form joint ventures is an attempt to share risk, but the main purpose here may be to access new sources of capital and technology rather than risk reduction per se.

Expectations Concerning Wheat Prices

Each manager was asked, using the visual impact method (Hardaker, Huirne, and Anderson 1997, 35–37), to specify subjective probability distributions of nominal prices for "their average quality" wheat for four dates in the future. The interviews were carried out toward the end of the growing season in 1998. Scenarios were framed such that the first two prices referred to the current crop, shortly after harvest and later in the selling season. The second two prices referred to the 1999 wheat crop, about which much less would be known, but important in that these prices form the basis for planting decisions in September 1998. Inflation was expected by the managers to continue at 60–70 percent per annum. Means and coefficients of variation of these distributions are reported in table 7.3.

While the means of the subjectively estimated probability distributions of wheat prices all show an increase over the period, the effect appears to be primarily an inflationary one. They do, however, imply a wheat price of over $180 per tonne in October 1999 at estimated exchange rates (compared to a forecast world price from the FAO of $169 per tonne). A feature of these estimates is the small coefficients of variation, which are well below values calculated from historical series. If these estimates of variability are an accurate representation of managers' views, then their low values raise questions. For the managers to make decisions on such values, it follows that their risk premia will be correspondingly small. Eliminating price risk would therefore make little difference to their resource allocation for the forthcoming crop. However, if these values do not represent the true picture (i.e., are irrational expectations) then the consequences of unforeseen poor outcomes and their frequency will be that much more serious. It is not possible to know whether or not these estimates of variability are "rational". The historic data suggest not. What this may mean is that these managers are not fully informed about all the factors that affect the price of wheat and its potential variability. This suggests that an effective market information system would lessen the consequences of price outcomes, which fall outside the range anticipated by the farmers.

Table 7.3 Subjective Means and Coefficients of Variation for Future Nominal Wheat Prices

	lei/kg			
Date	**1 Oct. 98**	**1 Apr. 99**	**1 Oct. 99**	**1 Apr. 00**
CCSC Manager				
Mean	958	1,458	1,654	1,886
Coefficient of variation	0.07	0.05	0.04	0.06
AS (1) Manager				
Mean	1,001	1,357	1,807	1,957
Coefficient of variation	0.03	0.02	0.02	0.01
AS (2) Manager				
Mean	1,186	1,310	1,842	–
Coefficient of variation	0.07	0.05	0.06	–

Source: Interviews with managers, June 1998

A Farm Level Model of Risk

An alternative way of investigating the impact of risk is to estimate its effect on farm incomes rather than, as the previous section has done, to observe managers' expectations of price variability. Anderson (1989) has proposed a simple model whereby an approximate indication of the importance to a farm business of a specific risk can be obtained. While the original model investigated the impact of a new source of risk (Hardaker, Huirne and Anderson 1997), it is used here to look at the consequences of the elimination of an existing source of risk, namely, that associated with wheat production. By combining a theoretical model with assumptions about risk attitudes and including data relating to the cropping systems, the aim is to throw light on the impact of risk associated with wheat production on farm income on the case study operations.

If T is the existing overall return, together with all its risk; X is the return from one component (e.g., wheat) of the overall return together with its risk; and Y is the remaining risk in the absence of variability associated with that component, then the certainty equivalent (CE) of T, can be approximated by:

$$CE\,[T] = E\,[T]\,(1 - D),$$

where D is a proportional risk deduction given by,

$$D = RC\,[X]\,(0.5C\,[X]\,Z + QC\,[Y])$$

where

R = coefficient of relative risk aversion
$C\,[X]$ = coefficient of variation of X
$C\,[Y]$ = coefficient of variation of Y
Z = relative size of the new enterprise in terms of expected outcomes
Q = correlation coefficient between X and Y

The model requires a number of assumptions about the variability in income from wheat and the other sources. The chosen levels for the variables were thought plausible, but the results need to be treated with caution. Furthermore, since the model assumes normality in the distributions, it will be subject to some of the concerns of Quiroz and Valdes (1997). The coefficient of variation of wheat income was set initially at 0.35, while the coefficient of variation of all other income was set at 0.3. The correlation coefficient of wheat income and other income was set at a relatively high 0.5. The coefficient of relative risk aversion was assumed to be at a "somewhat risk-averse (normal)" level of 1.0 (Hardaker, Huirne and Anderson 1997, 102).

It should be noted that the model relates to the elimination of all variability associated with wheat *income*, while a price stabilization scheme would only deal with price risk, leaving yield risk unaltered. Thus, a second set of calculations simulates a situation where the effect of the price stabilization scheme is to reduce the coefficient of variation of wheat income from 0.35 to 0.2.

Table 7.4 shows results for the four case study farms. If risk associated with wheat income were eliminated completely, the estimated increment to total income would be 7–8 percent for the CCSC and the ASs. There would, of course, be no gains for the smallholder who is not engaged in wheat production. The model is incapable of predicting farm system changes as a result of risk reduction and thus is not able to predict whether the smallholder would commence wheat production in the absence of risk. With the reduction in the coefficient of variation of wheat income from 0.35 to 0.2, the effect is a gain in total expected income of 3.5 percent, (i.e., 7.1–3.6) for the CCSC and 4 percent and 4.3 percent, respectively, for the two ASs.

Table 7.4 Increment to Expected Income (percentage) following Reduction of Risk in Wheat Production

	Income from Wheat (%)	Coefficient of Variation of Income from Wheat	
		0.35	0.2
	%	%	%
CCSC	31	7.1	3.6
AS (1)	41	7.8	3.8
AS (2)	50	8.3	4.0
Smallholder	0	0	0

Source: Case study farms and authors' calculations

If the chosen levels for the variables are plausible, the conclusion is that gains to these farms would be relatively small at less than 5 percent of total expected income. It could be argued, on the evidence of the managers' expectations concerning price variability and on the basis that the assumption concerning covariance may be relatively high, that the gains from reduction of anticipated variability may be considerably *lower* than those estimated.

In conclusion, the case studies suggest that Romanian farmers could avoid a good deal of risk through diversification, as soils are good and offer many cropping possibilities. Information on wheat prices and markets is in short supply and needs to be improved. A recurring problem is the shortage of working capital that suggests attention should be focused on providing an effective farm credit system. More importantly, the level of gains from the elimination of price risk in wheat does not suggest major benefits from improved resource allocation.

Conclusion

Provision of a safety net for cereal farmers has a long history in the United States through a loan-rate scheme and in the EU through intervention buying under the CAP. Safety-net support may lend credibility to the market system in Romania among farmers and policymakers not yet fully convinced of the reliability of market economics. All aspirant EU members, in their pre-accession phases, are tending to shadow and imitate the present CAP institutions and procedures. This suggests that if a market stabilization scheme is to be implemented in Romania, it will be based on the EU system of intervention buying. The cereals regime under the CAP was designed originally to provide both income support and price stability to European farmers, but Agenda 2000 makes clear that intervention in the future will offer a safety net at a much lower level than previously. Indeed, its role will diminish as EU cereal prices converge on world prices, the latter becoming more stable as agricultural trade becomes more liberalized.

An intervention system, however modified, may be required for eventual EU membership, but is it, or any other scheme, needed at the present time to stabilize wheat prices in Romania? Although reduction of price risk may be desirable, it is suggested that it is not an urgent priority for improvement of farm level decision-making in Romania. Even a scheme which is self-financing, and therefore requiring no state funds, will use scarce program management skills that at the present time would be better employed elsewhere. Elimination of market imperfections in the agri-food chain, particularly with respect to finance, grain storage, and market information, will give more immediate gains. Shortage of farm working capital is a major problem, and attention would be better directed, for example, at setting up an effective farm credit system. More fundamentally, international price transmission is the key to efficient resource allocation. Technical efficiency of wheat production will improve in response to international competition and world prices. Despite the attraction of price stabilization to policy-makers, the immediate aim in Romania should be to reduce market distortions and imperfections.

Acknowledgments

Adapted from Hubbard, L. J., Lingard, J., and Webster, J. P. G. 2000. Romanian wheat prices: Is there a need for stabilisation? *Food Policy* 25:55–67. Copyright 2000, with permission from Elsevier Science.

Notes

1. As a measure of the extent of price variability, the coefficient of variation (c.o.v.) expresses the standard deviation as a proportion of the mean. The higher the c.o.v., the greater the variability of price about its mean level.

References

Anderson, J. R. 1989. Reconsiderations on risk deductions in public project appraisal. *Australian Journal of Agricultural Economics* 33:136–140.

Hardaker, J. B., Huirne R. B. M., and Anderson J. R. 1997. *Coping with Risk in Agriculture*, London: CAB International, 274.

Hubbard, L. J., Lingard, J., and Webster J. P. G. 2000 Romanian wheat prices: is there a need for stabilization? *Food Policy* 25:55–67.

Quiroz, J. and Valdes, A. 1997. *On Agricultural Price Stabilization: A proposal for the transition.* Working Paper, World Bank.

Chapter 8

The International Competitiveness of Romanian Agriculture: A Quantitative Analysis

Matthew Gorton, Dinu Deaconescu, and Dan Petrovici

The aim of this chapter is to assess the international competitiveness of Romanian agriculture by means of a quantitative analysis of the distortions between input and output prices faced by domestic producers compared against equivalent international (social) prices. The ability of Romanian producers to operate profitably at social prices given 1997–1998 technical coefficients is assessed. Social profitability is an important indicator of a country's comparative advantage or disadvantage in the production of a particular good and in its ability to pursue opportunities to trade within a liberal international environment.

In this chapter, we describe three ratio measures of distortions and competitiveness and estimate these ratios by constructing a policy analysis matrix (PAM) for eight agricultural commodities. These commodities are chosen because of their importance in consumer purchases or the degree of support afforded to them by the present Romanian government and within the CAP of the EU. This methodology has been widely used in academic studies and in policy assessments by international development agencies. However, the present study is the first application of the PAM framework to Romanian agriculture.

Methodology for Assessing Distortions and International Competitiveness

An array of procedures and methods have been developed for estimating input and output distortions, trade and agricultural policy protection, and measuring the international competitiveness of a country's agricultural production. This chapter employs three indicators that are widely used in agricultural policy analysis: the nominal protection rate or coefficient, the effective protection rate, and the domestic resource cost ratio.

Nominal Protection Rate (NPR) or Coefficient (NPC)

The NPR is an indicator of the differential between domestic and international output prices for an agricultural commodity. Such differentials may be caused by trade policies (e.g., import tariffs, non-tariff barriers, government monopolies on trade), market imperfections, state agricultural policies, and exchange-rate policies. For commodity *i* (in percentage terms):

$$NPR_i = \left(\frac{P_i^d}{P_i^d} - 1 \right) \times 100 = \left(NPC_i - 1 \right) \times 100$$

where

P_i^d = the domestic price and

P_i^b = the border price (or world market reference price)

The border price is usually expressed in U.S. dollars, while the domestic price is converted using the market exchange rate. Border prices should be adjusted back to the farm level by accounting for transport, storage, port, and handling costs. A positive NPR, when the domestic price is higher than the border price, means that the output of the producer is nominally protected and the consumer is taxed. A negative NPR, when the domestic price is lower than the border price, shows that the producer is nominally taxed and the consumer is subsidized. Positive, negative, and zero NPRs correspond to NPCs greater than one, less than one, and equal to one, respectively. The most extensive use of NPRs for Central and

Eastern Europe has been by Bojnec and Swinnen (1997), albeit without adjusting for transport and storage costs.

Effective Protection Rate (EPR)

The EPR is an indicator for measuring trade-, price- and exchange-rate-related distortions. The EPR captures transfers due to distortions in input as well as output price on the product's value added, that is, output price (gross value) less specified (usually variable) traded input costs. The EPR for commodity *i* is defined (in percentage terms) as:

$$EPR_i = \frac{VA_i^d - VA_i^b}{VA_i^b} \times 100 = \left(\frac{VA_i^d}{VA_i^b} - 1 \right) \times 100$$

where

VA_i^d = the value added of output *i* at domestic prices and

VA_i^b = the value added of output *i* at border prices

The EPR can be positive, negative, or zero. A positive EPR indicates that the value added at domestic prices is higher than value added at border prices, and hence output *i* is "effectively" protected through the combination of domestic output and input price policy. By contrast, a negative EPR implies overall producer taxation: domestic value added is effectively taxed. When EPR is zero, the output *i* is neither taxed nor subsidized, and value added at domestic prices is equal to the value added at border prices.

Domestic Resource Costs (DRCs)

The DRC ratio also measures the relative efficiency of domestic production in terms of its international cost competitiveness. The DRC coefficient compares the opportunity costs of using domestic primary resources—land, labor, and capital—and of traded inputs to the value added by these resources evaluated at border prices:

$$DRC_i = \frac{j = \sum_{k=1}^{n} a_{ij} V_j}{P_i^b - \sum_{j=1}^{k} a_{ij} P_j^b}$$

where

$a_{ij} (j = k + 1 \text{ to } n)$ = the technical coefficient (input use per unit of output) for domestic resource (non-traded intermediary input) j in the production of output i,

V_j = the shadow price of such an input

When the DRC ratio is smaller than one, domestic production is efficient and internationally competitive because the opportunity cost of spent domestic resources is smaller than the net foreign exchange gained in export or saved by substituting for imports. A DRC ratio of less than one is thus taken as an indicator of long-run comparative advantage. The opposite is true when the DRC ratio is larger than one (Yao 1997).

These three indicators are often derived from a PAM framework. The PAM is based on the simple formula of profits equaling revenue minus costs. To construct the matrix, costs are disaggregated into tradable inputs and non-tradable inputs (often referred to as domestic resources or factors). Profits, revenues, and the two types of costs are then calculated using both actual prices (private prices) and social (or "efficiency") prices (FAO 1992). For tradable inputs and output, social prices are measured as adjusted border prices. For domestic factors (land, labor, and capital) for which no world market price exists, social prices are based on their domestic opportunity costs (i.e., what these factors would earn in their next best use, with the output of that use measured at border prices). The differences between private and social costs, revenues, and profits are transfers whose magnitudes reflect the extent to which domestic prices diverge from efficiency prices.

Table 8.1 Generic PAM Framework

	Revenue	Tradable Inputs	Domestic Resources	Profits
At Private Prices	A	B	C	D
At Social Prices	E	F	G	H
Transfers	I	J	K	L

The structure of a PAM is indicated in table 8.1. Within the PAM framework, the Nominal Protection Coefficient (NPC) can be calculated from table 8.1 as cell A divided by cell E. The three protection ratios can be expressed as:

Nominal Protection Coefficient NPC = A/E

Effective Protection Rate EPR = (A – B)/(E – F)

Domestic Resource Costs DRC = G/(E – F)

The PAM accounting framework is a consistent means of tabulating information required for price policy analysis, and its construction aids understanding of the overall magnitude of distortions in the production of agricultural commodities. Its main limitation is that it is constructed for a specific year and thus needs to be altered as principal parameters (such as world prices of outputs and inputs and yields) change over time. In some instances, even comparatively small changes in world prices may be highly significant in the assessment of whether a particular activity is socially profitable or unprofitable. However, once the PAM framework and component spreadsheets are constructed, parameter changes may be made relatively easily.

Data Collection, Sources, and Assumptions

The competitiveness of Romanian agricultural production was assessed for eight main commodities (wheat, maize, sugar beet, sunflower, milk, beef, chicken and pork meat) and two markets (world and domestic). The estimation of NPCs, EPRs, and DRCs utilized a number of data sources. These included: a survey of Romanian private farming (MoAF/World Bank/EC 1997); information on the Romanian voucher scheme for agricultural inputs (MoAF/Institute of Agricultural Economics 1998); National Commission for Statistics (1998) census returns; MoAF unpublished data (1998); and industry estimates.

Output Prices

Social prices for outputs and tradable inputs were taken as border prices (export/import parity prices), adjusted to the farm level by port and handling charges, transport, storage, and maintenance costs (where appropriate). Drying, storage, handling, and custom charges were given by Cargill representatives based in Romania during May 1998. For products for which Romania is a net exporter, an average free on board (f.o.b.) export parity price was taken as the unadjusted reference price. For products for which Romania is a net importer, average cost, insurance, freight (c.i.f.) import parity prices were used. Since Romania has oscillated between being a net exporter and a net importer of wheat, an average of the adjusted export and import parity prices for that commodity was taken in order to reflect trading patterns and conditions better.

Input Prices

Private input prices and quantities together with information on yields were taken from two surveys (MoAF/World Bank/EC 1997; MoAF/Institute of Agricultural Economics 1998). All the private prices used were average national prices, and no substantial analysis on a regional or subregional level was made.

Labor costs were based on lei values at the end of 1997, with different levels for skilled and unskilled labor. The social price of land was measured as its rental value in the most profitable alternative agricultural use. For example, if maize production represented the only alternative to wheat production, the social cost of land for the wheat activity was represented by the social profits (excluding land) from the production of maize (Monke and Pearson 1989). However, a single clear alternative is often not evident as systems vary in terms of riskiness and the desirability of crop rotation. For example, vegetable crops often provide higher returns on average than staple food crops, yet many producers continue to grow food crops because of their greater price stability from year to year. Following Pearson, Avillez, and Bentley (1987), an average of suitable commodity alternatives was taken for deriving shadow land prices, as social land values would probably fall somewhere in this range.

Farm Structure

For the four crops analyzed (wheat, maize, sugar beet, and sunflower), the DRCs were estimated only for production by ASs due to the lack of reliable data for other management types. It is hoped that this analysis will be replicated for other organizational types, as such disaggregation would take into account the wide variations in input use, yields, and other production costs between farm types. The DRCs were estimated for nonirrigated land. The Romanian system of irrigation is highly inefficient and in a poor state of repair (OECD 2000). Only 4 percent of private farmers in the MoAF/World Bank/EC (1997) sample made use of such services, and the mean yields of irrigated and nonirrigated land did not differ significantly. The social value of water in Romania is difficult to estimate, and given the small number of farmers using irrigation services it was not possible to construct robust estimates of private and social revenues and costs of irrigated farming systems by individual commodity.

The technical coefficients used in the construction of the PAMs for pork and chicken were based on figures for the large state and former state complexes, which still dominate production for these commodities in Romania. Coefficients for the household livestock production, which is mainly for self-consumption, are likely to differ substantially. The technical information was supplied by MoAF officials and was checked where possible with industry representatives (e.g., Union of Poultry Producers).

Time Period and Debt Treatment

The PAM coefficients were estimated for the period 1997–1998. Prior to the two surveys, complete data on tradable and non-tradable input use by commodity was not available for Romania. As a result, the evolution of competitiveness as measured by DRCs cannot be estimated for previous years, although it is hoped that the analysis presented here will provide a basis for future research. In estimating private and social profitability in the period 1997–1998, no adjustment was made for the debts of producers and the costs of debt servicing from previous time periods—the usual approach in estimating DRCs. However, while in a particular time period an enterprise's activities may be privately and socially profitable, the income generated may not be sufficient to cover the costs of debt servicing and repayment. In Romania, the debts accumulated by certain enterprises are highly significant and pose a severe threat to their long-run viability.

Sensitivity Analysis and Marginal Production

Sensitivity analysis provides a way of assessing the impact of changed assumptions in estimating both private and social profitability. The value of the Romanian lei has been unstable during the transitional period (with a large nominal depreciation), and changes in the exchange rate receive most attention in this analysis. The closer the DRC is to 1, the more marginal is Romania's comparative advantage or disadvantage in the production of that particular commodity, and enterprises in this range are most susceptible to changes in world market prices and exchange rates. By contrast, the further the DRC is below 1, the more robust the level of international comparative advantage. Finally, it should be noted that the DRCs estimated here are sector or production-type averages. Some producers, especially those entering or leaving the sector as marginal enterprises, may be considerably more or less efficient than the sector average and so vary in their susceptibility to adverse changes in international prices or exchange rate shifts.

Results

Arable Sector

Tables 8.2, 8.3, 8.4, and 8.5 present the PAMs for wheat, maize, sugar beet, and sunflower, respectively, assuming yields of 3, 4.2, 27, and 1.4 tonnes per hectare for bread wheat, maize, sugar beet, and sunflower, respectively, as grown by ASs. These figures are based on farm-type average yields provided by the MoAF. The social and private values are shown per tonne and can be grossed up to provide the magnitude of total transfers for each production system in Romania. The three competitiveness ratios (NPC, EPR, and DRC) are displayed as key summary ratios.

The Romanian farm-gate price for wheat lies between the adjusted c.i.f. and f.o.b. border prices, and, using international prices at the beginning of 1998, Romanian wheat production is both socially and privately profitable (table 8.2). However, the sharp downturn in international wheat prices since that date will have reduced Romania's comparative advantage and profitability. This instance highlights the sensitivity of the analysis (and of farming systems with low tariff and budgetary protection) to changes in external parameters.

Table 8.2 PAM for Romanian Wheat Production by Agricultural Societies, 1997–1998 (lei per tonne)

	NPC	EPR	DRC	
Summary Ratios	1.25	1.25	0.78	
	Revenue	**Tradable Costs**	**Domestic Factors**	**Profits**
Private	950,000	550,382	297,038	102,580
Social	947,375	627,310	249,346	70,719
Transfers	2,625	–76,928	47,692	31,861

Romanian maize production is both privately and socially profitable, albeit modestly, when the voucher scheme is taken into account (table 8.3). Farmers are effectively protected by the domestic price being above the adjusted border price (NPC of 1.1), but are penalized by tradable and non-tradable costs being above social prices. The DRC, while indicating international comparative advantage, is nonetheless close to 1, and competitiveness will be sensitive to changes in world prices.

In contrast, sugar beet production is not profitable in either private or social terms. Again, while output prices are above equivalent parity prices, domestic tradable and non-tradable costs are above their social rates (table 8.4). The DRC is significantly above 1, and it is difficult to envisage the bulk of sugar beet production in Romania being internationally competitive.

Table 8.3 PAM for Romanian Maize Production by Agricultural Societies, 1997–1998 (lei per tonne)

	NPC	EPR	DRC	
Summary Ratios	1.10	1.16	0.90	
	Revenue	**Tradable Costs**	**Domestic Factors**	**Profits**
Private	700,000	403,789	287,444	8,767
Social	635,439	380,271	230,388	24,780
Transfers	64,561	23,518	57,056	–16,013

Table 8.4 PAM for Romanian Sugar Beet Production by Agricultural Societies, 1997–1998 (lei per tonne)

	NPC	EPR	DRC	
Summary Ratios	1.16	1.50	2.41	
	Revenue	**Tradable Costs**	**Domestic Factors**	**Profits**
Private	220,000	145,037	160,016	–85,053
Social	188,993	139,180	120,212	–70,399
Transfers	31,007	5,857	39,804	–14,654

Table 8.5 PAM for Romanian Sunflower Production by Agricultural Societies, 1997–1998 (lei per tonne)

	NPC	EPR	DRC	
Summary Ratios	1.04	1.00	0.80	
	Revenue	**Tradable Costs**	**Domestic Factors**	**Profits**
Private	2,100,000	1,010,679	975,199	114,122
Social	2,023,454	936,304	869,352	217,798
Transfers	76,546	74,375	105,847	–103,676

The oilseeds sector is more promising. Sunflower production is only marginally protected (NPC = 1.04) and is profitable by both private and social standards (table 8.5). Moreover, in international terms, of the four crops analyzed, sunflower production is easily the most attractive. Romania's agro-climatic environment is conducive to this crop and, despite very low input use, internationally respectable yields are derived.

An attempt has been made in the case of wheat to adjust the private tradable costs for the effects of the Romanian voucher scheme. The scheme, which is discussed in chapter 2, was designed to increase the share of input supply by the private sector and provide transitional relief to farmers. Quantifying the impact of the voucher scheme is complex. First, where farmers both own private plots and are members of ASs, it is difficult to know which farm type benefits from the vouchers. Second, it is hard to know the breakdown of voucher use by crop type, especially given that many inputs are lumpy. A MoAF survey of 384 small landowning households (average holding size of 2.75 ha) on the impact of the voucher scheme found that 47 percent of households spent their vouchers on fertilizers, 27 percent on mechanical services, 19 percent on diesel oil, 9 percent on seeds, and 2 percent on pesticides (Luca 1998). If one assumes that the proportion of vouchers used in wheat production mirrors its overall share of crop production, the scheme could have lowered input costs by over 30 percent (table 8.6). In the absence of the voucher scheme, domestic tradable costs would have been above international prices.

Considering the two main aims of the voucher scheme—enhancing the role of the private sector and competitive markets in input provision, and the granting of transitional relief—the results of its implementation have been mixed. Input supply is still dominated by state-owned companies for fertilizers and seeds, for example, Unisem, Semrom, and Comcereals, and the hire of agricultural machinery by Agromechs; the privatization of which slowed considerably in 1998–1999. Furthermore, these input suppliers have benefited not only by being the main recipient of vouchers but also from other direct budgetary support. In 1997, the MoAF budget set aside lei 200 billion for fertilizers, lei 274 billion for irrigation and drainage, and lei 149 billion for certified seed subsidies. However, the domestic prices of tradable inputs such as fertilizers were still above world market levels. In this analysis, the NPCs for tradable inputs are all greater than 1, and suppliers are protected by tariffs of up to 20 percent. If one adjusts for quality differentials, the magnitude of protection is larger.

Table 8.6 Subsidy Equivalent of the Voucher Scheme for Wheat Production for Agricultural Societies (per ha)

Product	Quantity	Quantity Covered by Subsidy	Quantity without Subsidy	Price (lei)	Total Value	Subsidy	Non-subsidy Value
Certified seed	241	137	104	1,979	476,939	271,123	205,816
Nitrogen fertilizer	70	50	20	1,329	93,030	66,450	26,580
Complex fertilizer	229	0	229	1,612	369,148	0	369,148
Fuel	219	108	111	2,223	486,837	240,084	246,753
Plowing				146,250	229,294	146,250	83,044
Rotivation				148,770	148,770	0	148,770
Application of land treatments			299,219	299,219	0	299,219	
Sowing				113,306	113,306	63,680	49,626
Harvesting				247,530	247,530	0	247,530
Total					2,464,073	787,587	1,676,486
Subsidy Equivalent						32%	

There has thus been an asymmetry in treatment between agriculture and its input industries in Romania. While, as part of the World Bank Agricultural Structural Adjustment Loan (ASAL), the former faced tariff reductions and the removal of price supports, the tariffs on inputs were not cut and domestic input suppliers have benefited from both direct subsidies and the voucher scheme. In some cases the voucher scheme only partially offset the higher prices for inputs. This distortion handicapped the agricultural sector while the input supply industry remained heavily protected. It is important for policymakers to realize that if one wants a low-subsidy, internationally competitive agricultural sector, farmers should not be penalized through high input prices. The implementation of the ASAL has not, as yet, stimulated an international competitive input industry or allowed farmers access to externally produced inputs at world market prices.

Figure 8.1 charts the evolution of cropped areas in Romania between 1990 and 1999. As part of developing an internationally competitive sector, one would expect to see a specialization in those commodities for which Romania has a comparative advantage. This has in part occurred; the cropped area of oilseeds has increased since the end of Communism, and the area of land under sugar beet has diminished. The wheat cropped area has been fairly unstable, and this in part reflects the frequent changes in policy intervention (and the incentives they presented) by previous governments (Gorton et al. 1998).

Figure 8.1 Evolution of Cropped Areas in Romania, 1992–1999 ('000 ha)

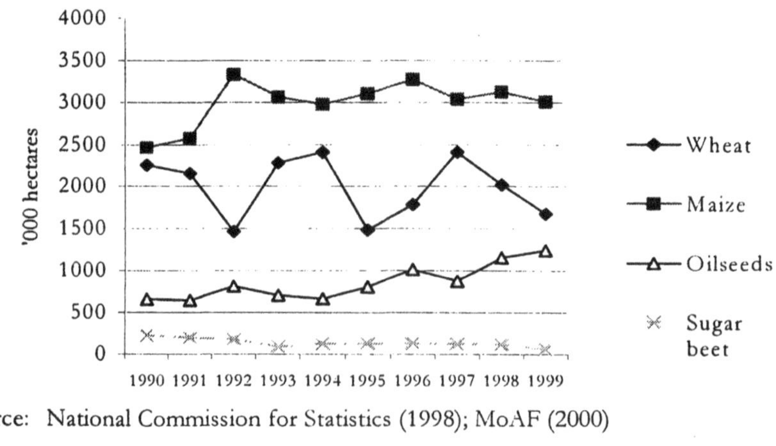

Source: National Commission for Statistics (1998); MoAF (2000)

Livestock Sector

The competitiveness measures for the livestock sector are presented in tables 8.7–8.10. Overall, Romanian meat and milk production is less competitive than crops by international standards. Livestock NPCs are higher than for the crops analyzed, and levels of effective protection are greater. Only for pork is the DRC estimate below 1 and in this case only marginally (0.97).

Table 8.7 PAM for Romanian Pork Production, 1997–1998 (lei per tonne)

	NPC	EPR	DRC	
Summary Ratios	1.27	1.34	0.97	
	Revenue	**Tradable Costs**	**Domestic Factors**	**Profits**
Private	12,000,000	7,977,953	3,226,796	795,251
Social	9,451,466	6,457,995	2,889,858	103,613
Transfers	2,548,534	1,519,958	336,938	691,638

The technical coefficients used in the competitiveness analysis of pork production (table 8.7) are based on figures for the large state and former state complexes, referred to here as CCSCs, that dominate marketed production in this sector. The largest share, approximately 60 percent, of domestically marketed pig production comes from these specialized state and former state-owned complexes. Many of these enterprises are vertically integrated and incorporate breeding activities, slaughtering, processing, and sometimes retail shops. There were 55 large-scale pork complexes in Romania, of which, at the end of June 1998, twelve were privatized, twenty-four were proposed for privatization by the end of the year, seven were in administrative liquidation, and twelve were in judicial liquidation (MoAF 1998).

The calculations are based on a feed conversion ratio of 4.8 (4.8 kg of feed to 1 kg of meat gained), and a mortality rate of 5 percent. For comparison, the UK figures are lower: feed ratios of 2.59 for rearing and 1.79 for breeding, and a mortality rate of 2.5 percent for rearing and 4.4 percent for breeding (MLC 1997).

Chicken production is effectively protected in Romania. Output prices are above adjusted border prices (NPC = 1.23) and profits at social prices would be negative (table 8.8). As with the pig industry, poultry production is dominated by large state and former state-owned enterprises. Fifty-eight large poultry complexes dominated the urban market in 1997, and of these, fourteen were privatized, nineteen proposed for privatization, eleven in administrative liquidation, and the remaining fourteen in judicial administration. A feed conversion ratio of 2.6 and mortality rate of 10 percent were used in this analysis. These figures were obtained from ministry officials and checked by a representative from the Union of Poultry Breeders. Private tradable input costs are higher than if measured at social prices, largely because of higher feed costs.

Table 8.8 PAM for Romanian Chicken Production, 1997–1998 (lei per tonne)

	NPC	EPR	DRC	
Summary Ratios	1.23	1.29	1.20	
	Revenue	**Tradable Costs**	**Domestic Factors**	**Profits**
Private	11,700,000	6,900,245	2,856,928	1,942,827
Social	9,494,500	5,786,534	4,439,364	–731,398
Transfers	2,205,500	1,113,711	–1,582,436	2,674,225

Unlike pig and poultry, small-scale, fragmented producers dominate beef and milk production in Romania. Both beef and milk production are uncompetitive at international prices (tables 8.9 and 8.10, respectively).

Table 8.9 PAM for Romanian Beef Production, 1997–1998 (lei per tonne)

	NPC	EPR	DRC	
Summary Ratios	1.31	1.45	1.15	
	Revenue	**Tradable Costs**	**Domestic Factors**	**Profits**
Private	11,200,000	6,835,424	3,625,572	739,004
Social	8,542,772	5,528,367	3,478,421	–464,016
Transfers	2,657,228	1,307,057	147,151	1,203,020

The NPC for beef is comparatively high (1.31), with a markup of over lei 2.5 million per tonne of liveweight beef between domestic and international prices. A feed conversion rate of fifteen and a mortality rate of 5 percent were assumed for the beef calculations.

Tradable costs are again higher on the domestic market because of higher feed costs.

Table 8.10 PAM for Romanian Milk Production, 1997–1998 (lei per hectoliter)

	NPC	EPR	DRC	
Summary Ratios	1.28	1.31	1.15	
	Revenue	**Tradable Costs**	**Domestic Factors**	**Profits**
Private	3,000,000	719,540	1,334,790	945,670
Social	2,342,600	599,290	2,012,900	−269,590
Transfers	657,400	120,250	−678,110	1,215,260

Romanian milk prices are higher than their international equivalents and the sector is not competitive by international standards (table 8.10). This is readily understandable given an average herd size of less than two cows, with high transaction costs, unrealized economies of scale, and an erratic and poor quality of supply. Less than 20 percent of the milk produced in Romania is delivered to dairies with the rest consumed within the household. The milk that is sold to dairies by individual farmers tends to have high bacterial cell counts due to the lack of proper cooling facilities, and in some cases the poor quality of the raw material renders it only suitable for manufacturing cheese.

The small scale of private agriculture in Romanian is highlighted in the results of the MoAF/World Bank/EC (1997) survey, which found that the average size of milking herds was 1.4; the comparable figure for pigs was 2.3. The largest private milking herd was twenty-six—tiny by international standards.

The calculations for all the livestock products analyzed allow for credit costs, based on the provision of working capital, but do not include the cost of meeting the interest on previous debts. For the state and former state complexes, these debts are considerable (table 8.11). At the end of 1997, this total level of debt in the pig industry was lei 3,120 billion with interest costs of over lei 1,311 billion. These costs account for the net loss made by the sector in that year (lei 663 billion). These debts are a major disincentive to potential new foreign and domestic investors. If they could be removed, the sector would be both privately and socially profitable, albeit in the latter case, only marginally.

Romanian livestock inventories fell sharply during the 1990s (table 8.12). The decline in livestock output has been greater than for arable commodities and this is consistent with the DRC estimates, which indicates that the latter are more internationally competitive. Regarding livestock, pork production and poultry output by large production complexes have witnessed the greatest declines. During the communist era, the share of total output accounted for by small household producers was much higher for milk than poultry or pork production. For example, in 1986—1989 household producers accounted for 58 percent of total milk output (OECD 2000). Meat and milk production by these small-scale producers has not witnessed the same scale of output decline as that of the large state and former state-owned livestock complexes.

Table 8.11 Financial Performance in the Romanian Pig Industry (billion lei)

Form of Enterprise	Total Share Capital	Indebtedness				Net (+) Profit/(–) Loss	
		31.12.96		31.12.97		31.12.96	31.12.97
		Total	Interest	Total	Interest		
Total, of which:	1,836.74	2,140.39	381.55	3,120.34	1,311.54	–150.59	–663.16
Privatized CCSCs	269.12	366.54	64.08	528.77	195.76	–38.45	–146.60
Admin. liq. CCSCs	86.71	84.14	15.28	118.41	54.88	–22.98	–81.07
Jur. liq. CCSCs	129.33	81.03	15.66	99.23	9.45	–10.82	–35.15

Admin. liq. = administrative liquidation
Jur. liq. = juridical liquidation
Source: MoAF (1998)

Table 8.12 Romanian Livestock Inventories, 1990–1999 ('000 head)

	1990	1992	1994	1996	1998	1999	1999/90 (%)
Pigs	12,003	9,852	7,758	8,235	7,194	5,749	47.9
Poultry	121,379	87,725	70,157	78,478	69,480	68,090	56.1
Cattle (beef/veal)	5,381	3,683	3,481	3,435	3,143	3,038	56.5
Milking cows	2,123	2,025	1,963	1,939	1,790	1,756	82.7

Source: OECD (2000)

International Comparisons

The DRC results for Romania can be compared with similar studies of international competitiveness in the region, although one must be sensitive to the different assumptions used (Gorton and Davidova 2001). The results presented in table 8.13 highlight that crop production in the region has generally been more internationally competitive than livestock farming.

Combining the 8 DRC ratios estimated for Romania with the 6 studies quoted in table 8.13, 114 cases are presented. From these 114 cases, there are 51 instances of DRCs of less than 1 (internationally price competitive). If one compares livestock with arable production, a clear distinction is evident. Out of fifty-one cases in the livestock sector, only six instances of DRCs below 1 are reported. In contrast, forty-five cases of international competitive arable production are reported (out of sixty-three cases). Within the arable sector, the most competitive crops were wheat and oilseeds (rapeseed in the north and sunflower in the south). Wheat producers benefited from comparatively high international prices during the period analyzed, especially in 1996–1997. After 1996, the pattern has been less pronounced as world arable prices fell. This fall reduced the measured competitiveness of CEECs' grain production by making the traded output less valuable, and increased the competitiveness of livestock production by reducing the traded input cost of fodder. Looking at production patterns and international competitiveness, while input and output intensity has decreased during transition, the fall in input use has been greater (Banse et al. 1999). Labor input has fallen as, typically, employees who left former state and cooperative farms have not been replaced (OECD 1999).

Table 8.13 DRC Calculations for CEEC Agriculture by Commodity

	Bulgaria			Czech Republic			Hungary			Slovakia			Slovenia	Poland		
Year	1994	1995	1996	1994	1995	1996	1994	1995	1996	1994	1995	1996	1998	1996	1997	1998
Wheat	0.45	0.32	0.26	0.82	0.49	0.47	0.64	0.71	0.89	0.76	0.47	0.33	1.47	0.53	0.81	0.99
Maize							1.28	0.90	0.82	1.05	0.70	0.35	1.62			
Barley	0.80	0.70	0.41	1.49	0.76	0.54	1.21	0.77	1.27	2.27	0.75	0.22				
Sunflower							0.76	0.72	0.80	0.77	0.59	0.56				
Sugar beet										0.63	0.69	0.99		3.38	2.70	4.08
Rye										1.37	2.11	0.43		1.07	1.25	2.45
Rapeseed										0.44	0.80	1.04		0.73	0.61	0.73
Milk	1.40	1.68	1.15	2.54	1.89	1.96	6.90	6.13	13.98	5.48	2.85	2.84	1.37	1.66	2.04	3.32
Beef	1.61	1.38	0.35	2.16	1.76	1.93	1.31	2.53	2.53	–4.00	–6.20	11.90		1.18	2.95	3.71
Pork	0.92	0.68	0.64	3.10	1.74	1.40	5.41	–3.84	2.88	–0.70	3.79	0.67	1.21	1.65	1.78	2.14

Source: Ratinger (1997); Bozik et al. (1998); Banse et al. (1999); Kuhar (1999); Gorton, Davidova, and Ratinger (2000); and Gorton et al., (2001)

The only country that does not record any instances of internationally competitive crop production is Slovenia. However, the analysis suffers from the paucity of data, although the figures for Slovenia are consistent with the results of other measures of competitiveness, which reveal domestic production to be highly protected and predominantly noncompetitive on international markets (Sarris 1998). Agricultural production in Slovenia is hampered by unfavorable natural conditions, fragmented farm structures, and the relatively high costs of domestic factors (especially labor and capital).

At first glance, Bulgaria appears to be the most competitive arable producer in the region. However, one should be careful in interpreting the results on two counts. First, the technical coefficients supplied by the Ministry of Agriculture, on which the Bulgarian analysis is based, are biased toward large farms. Second, during the analyzed period, the domestic prices for wheat and barley were kept below world market prices through the use of various impediments to export. In other words, farmers were effectively taxed in an environment of an unstable exchange rate and macroeconomic crisis. As the real cost of factors of production rises with economic recovery and distortions start to be removed, DRCs are likely to increase (Gorton, Davidova, and Ratinger 2000).

As in Romania, the international competitiveness of livestock production in the region as a whole tends to be much lower than for the arable sector. The DRCs for beef and milk production in every case reported were greater than 1. Relatively low international prices, high labor and feed use as well as disadvantageous farm structures have all contributed to this situation. In Hungary and the Czech Republic, the situation has been slightly better due to higher real incomes and the attraction of greater levels of foreign direct investment (FDI) in the dairy sector. In these countries foreign-owned dairies have provided investment credits and other support to larger dairy farmers to help ensure regular supplies (Gow and Swinnen 1999).

Conclusion

The PAM framework provides a baseline for measuring the direction and magnitude of divergences between the prices faced by domestic producers and international (efficiency) prices in the main subsectors of Romanian agriculture. This framework can be readily adapted to incorporate future changes in domestic margins, international prices, and technical coefficients. The calculation of DRCs allows the comparison of efficiency among systems that produce unlike outputs.

Domestic prices for the eight Romanian commodities are all higher than comparative adjusted border prices (NPCs equal to or greater than 1), indicating that farmers are effectively protected. However, the degree of protection for wheat, sunflowers, and maize is modest. The greatest output price protection is for milk and beef. Tradable input prices are, in general, also above international prices, reflecting the level of unreduced tariffs for these inputs and inefficiencies in domestic input supply industries.

At early 1998 international prices, DRCs are lower than 1 for the production of maize, sunflower, and wheat, indicating that these crops are socially profitable. Romanian farmers operating at social prices for output and inputs could make positive returns from these crops. In contrast, the DRC for sugar beet is greater than 1, indicating that Romanian production for this crop is not internationally competitive. There has been some specialization in agricultural production according to international comparative advantage during the period of transition: the area devoted to sunflowers and maize has increased while the cropped area of sugar beet has significantly diminished. However, further opportunities for specialization exist, and Romania has the potential to expand its trade in sunflower products.

As the prices of tradable inputs used by Romanian farmers are still above international prices overall, agricultural producers would benefit from a reduction in tariffs on tradable inputs and the completion of privatization of seed, fertilizer, and feed milling enterprises. If farmers are expected to operate in a low-subsidy environment without price support, it is essential that inputs are available at internationally competitive prices.

International prices for fertilizers are lower than those charged by Romanian producers and there are strong arguments for lowering the tariffs on traded agricultural inputs. By removing these trade restrictions and opening up the market, the price of agrochemical inputs should fall. This would benefit the beleaguered farming sector that has been squeezed by input prices rising at a faster rate than output prices. Of all agricultural inputs and services, fertilizers are the closest to being pure private goods. They are both subtractable and excludable, and as such are an obvious candidate for private supply (Carney 1998).

Milk production in Romania is not internationally competitive, which is readily understandable given an average herd size of 1.4 cows with high transaction costs, unrealized economies of scale, and an erratic and poor quality of supply. Even with the removal of price and budgetary supports, domestic milk prices are high by international standards. With a vicious circle of low retained earnings and suboptimal asset bases, this problem could persist for a considerable time without assistance. The development of assisted leasing arrangements may be a method for increasing dairy herd sizes and ensuring that cows go to farmers who will farm them most efficiently.

Romanian beef and chicken meat production is also currently not competitive when measured in social prices. The outlook for pork production is slightly brighter. However, the costs of credit and debt servicing have had a major impact on the profitability of former and present CCSCs producing pork and poultry. Sustained efforts to clarify the debts of state enterprises and to reduce accumulated debts are an essential element for improving the privatization process. If these inherited debts were to be removed for pork-producing complexes, the analysis conducted here indicates that these enterprises could be socially profitable, albeit modestly. However, upon accession to the EU, Romania will face strong competition for pork and poultry, for which there is no internal price support in the EU.

References

Banse, M., Gorton, M., Hartell, J., Hughes, G., Köckler, J., Möllman, T., and Munch, W. 1999. The evolution of competitiveness in Hungarian agriculture: from transition to accession. *MOCT-MOST: Economic Policy in Transitional Economies* 9:(3) 307–318.

Bojnec, S. and Swinnen, J. F. M. 1997. The pattern of agricultural distortions in Central and Eastern Europe. *Food Policy* 22:289–306.

Bozik, M., Blaas, G., Morrison, J., Hajnovicova, V., Lapisakova, J., and Mokrasova, V. 1998. *The Role of the Agriculture Sector in the Transition: Slovak Case Study*. FAO-ESA Project, Bratislava.

Carney, D. 1998. *Changing Public and Private Roles in Agricultural Service Provision*. London: Overseas Development Institute.

Food and Agriculture Organization (FAO). 1992. Agricultural price policy: government and the market. *Training Materials for Agricultural Planning* No. 31. Rome: FAO

Gorton, M, Daniłowska, A., Jarka, S., Straszewski, S., and Zawojska, A. 2001. The international competitiveness of Polish agriculture. *Post Communist Economies* 11:(4) 445–457

Gorton, M. and Davidova, S. 2001. The international competitiveness of CEEC agriculture. *The World Economy* 24:(2) 185–200.

Gorton, M., Davidova, S., and Ratinger, T. 2000. The competitiveness of agriculture in Bulgaria and the Czech Republic vis-à-vis the European Union. *Comparative Economic Studies* 42:(1) 59–86.

Gorton, M., Esanu, C., Deaconescu, D., Buckwell, A., and Davidova, S. 1998. *Impediments to Efficiency in the Agri-food Chain: Romania*. Report to the Organization for Economic Co-operation and Development (OECD), Paris.

Gow, H. and Swinnen, J. F. M. 1999. Impact of foreign direct investment on agriculture and agro-industry in transition economies. Paper presented to the *Second World Bank EU Accession Workshop in the Rural Sector*, in Warsaw, Poland 27–29 June.

Kuhar, A. 1999. *Application of the Policy Analysis Matrix on four agricultural production systems in Slovenia*. MSc. Dissertation. Wye College, University of London.

Luca, L. 1998. *Instruments of Agricultural Policies in 1997*. Bucharest, Romania: Ministry of Agriculture and Food.

Meat and Livestock Commission. 1997. *1997 Pig Yearbook*. Milton Keynes, UK: Meat and Livestock Commission.

Ministry of Agriculture and Food. 1998. *Farm Accounting Data*. Bucharest, Romania, unpublished.

———. 2000. *Romanian Agricultural Monitoring Report*. Bucharest, Romania. Unpublished.

Ministry of Agriculture and Food/World Bank/European Commission. 1997. *Private Agriculture in Romania: 1996 Farm Survey*, Bucharest, Romania. Mimeo-graphed.

Ministry of Agriculture and Food/Institute of Agricultural Economics. 1998. *Evolution sectorului alimentar in Romania: Raport annual 1997 al Ministerlui Agriculturi si Alimentaiei (The evolution of the agri-food sector in*

Romania: Annual report of the Ministry of Agriculture and Food). Bucharest, Romania: Ministry of Agriculture and Food/Institute of Agricultural Economics.

Monke, E. A. and Pearson, S. R. 1989. *The Policy Analysis Matrix for Agricultural Development.* Ithaca, N.Y.: Cornell University Press.

National Commission for Statistics. 1998. *Anuarul statistic al României.* Bucharest, Romania: National Commission for Statistics.

Organization for Economic Cooperation and Development. 1999. *The Impacts of Policy Reform on Labor Adjustment in Agriculture*, Doc. CCNM/EMEF/CA(99)10. Paris: Organization for Economic Cooperation and Development.

———. 2000. *Review of Agricultural Policies: Romania*, Paris: Organization for Economic Cooperation and Development.

Pearson, S. R., Avillez, F., and Bentley J. W. 1987. *Portuguese Agriculture in Transition.* Ithaca, N.Y.: Cornell University Press.

Ratinger, T. 1997. *The Competitiveness of Czech Agricultural Producers in an Integrated European Market.* EU ACE Programme on Agriculture and East-West Integration. Final report VUZE, Prague.

Sarris, A. H. 1998. *Policy Options for Slovenian Agriculture in an EU Accession Environment*, FAO Technical Co-operation Programme. Final report (Budapest).

Yao, S. 1997. Comparative advantages and crop diversification: A policy analysis matrix for Thai agriculture. *Journal of Agricultural Economics* 48:211–222.

Chapter 9

The Impact of the Association Agreement between Romania and the EU on Romanian Agriculture

Marc Duponcel[1]

The Association Agreements, or Europe Agreements (EAs), as they are formally referred to, between CEECs and the EU cover political dialogue, economic integration, and cooperation in different spheres (financial, cultural, etc.). These agreements aim at establishing free trade in industrial products. The first six associated countries, Bulgaria, the Czech Republic, Hungary, Poland, Romania, and Slovakia, have benefited from virtually completely free access to the EU market since 1995 (with the exception of the markets for some textile products). Trade liberalization in agriculture is more limited in scope; preferences are granted for limited quantities, and tariff barriers are maintained at a reduced level.

Romania signed an EA in February 1993. After ratification by the national parliaments, the agreement as a whole entered into force in February 1995. However, from May 1993 the framework of the so-called interim agreement allowed for trade provisions to be implemented.

As well as the EA, an agreement for wine was adopted in December 1993 and came into force on 1 January 1994. Similar agreements for wine were signed with Bulgaria and Hungary. They featured two main aspects. First, they agreed on reciprocal protection of the geographic indications and of the traditional names used to designate and present wines. Second, they granted preferential tariff quotas for wine products on a mutual basis. The analysis in this chapter includes the wine agreement.

The chapter focuses on a study of the effects of the EA on Romanian exports to the EU. It does not address the impacts of the EA on EU exports to Romania.

An Outline of the Agreement

When the EU negotiates preferential agricultural trade agreements with other countries, its usual practice is to use quantitatively limited preferences based on actual past trade. This practice was followed in the EAs, where preferences for all agricultural products were offered on the basis of actual past trade. In general, trade in the three years preceding the EA was taken into consideration. However, in the case of Romania's exports, it was agreed to take 1988 and 1989 as reference points because of the export ban introduced after the overthrow of Ceaucescu in 1989.

In the case of the countries in transition, this EU practice is highly questionable. Transition encompasses a systemic set of reforms, which is also likely to alter trade structures. Therefore, there is no reason to think that agri-food exports from the CEECs during transition would follow the same pattern as past trade. The prevailing opinion is that the pre-reform trade pattern of the CEECs was not the outcome of comparative advantage but was highly distorted by central planning. By freezing the past trade structures of the CEECs, which were surely not based on comparative advantages, the EU approach did not help the "natural" reorientation of past trade patterns. Designing preferences based on past trade was, therefore, equivalent to supporting the old trade commodity and country structure and disregarding the new export potential which would result from the implementation of the reforms. An additional feature for Romania, which might have worsened the picture, was that in the late 1980s it implemented a policy of forced exports to the western European countries in order to repay its external debt. The export ban of 1990 was an immediate reaction to this policy.

Initially, agricultural preferences in the EAs were negotiated for a period of five years. Before midterm, the preferences were re-negotiated for a second period, which started in mid-1997.

A detailed presentation of the main preferences granted by the EU is beyond the objectives of this chapter. A brief summary is given in box 9.1. For full details the reader is referred to the official text of the agreement.

Box 9.1 Summary of Trade Liberalization in EU Agricultural Imports as Foreseen in the EA

Asymmetry	The EU opens up its market faster and to a larger extent than the associated countries.
Two five-year periods	For Bulgaria, Hungary, Poland, the Czech Republic, Romania, and Slovakia preferences were originally set for five years and were then scheduled to be renegotiated.
Safeguard clauses	– One general safeguard clause under common provisions (preliminary notification to the joint committee necessary). – One safeguard clause specific for agricultural products without the necessity of bilateral consultation before implementation.
Two kinds of preferences granted by the EU	– Quantitatively unlimited preferences for some products (horse meat, some fruits and vegetables, etc.). – Quantitatively limited preferences for other products. For the first six associated countries the preference was 20 percent of levy in the first year and reached 60 percent from the third year on. The preference has been increased to 80 percent since 1995.
Preferences based on actual past trade	Most preferences have been granted on the basis of so-called traditional trade (in the case of Romania, trade during 1988–1989).
Annual quota increase	For the first five-year period the annual quota, within which the quantitatively limited preferences applied, increase was 10 percent of that of the base year. With the additional protocols the new growth rate has been 5 percent since mid-1997.
Preferences for non-Annex II products	Preferences granted by the EU for highly processed products (the so-called non-Annex II products) are included in separate protocols. Provisions differ from country to country.

The Main Features of the Preferences Granted by the EU

The preferences are granted in the form of preferential quotas and unlimited tariff rates. However, the level of utilization of the preferences depends on several features of those preferences, such as: the way the quotas are administered by the European Commission (EC); the extent of the preference rates; and the procedure used to negotiate the preferences. This section presents the major features of these preferences and their potential impact on the utilization by importers and exporters.

Preferential Quotas

Preferences granted for agricultural and processed products are the responsibility of two separate administrations in the EC. Trade in products subject to a common market organization under the CAP is administered directly by the Directorate General for Agriculture (DGAgri) of the EC, whereas trade in other products is administered by the Customs Administration. The main products included in the first group are beef, milk, pork, poultry and eggs, and cereals.

Preferential quotas administered by Customs are filled on a "first come, first served" system based on the customs declarations of the member states. When the quotas are filled, imports are carried out on a Most Favored Nation (MFN) basis. MFN levels of duties indicate the non-preferential tariffs charged for countries with which Romania has no trade agreement. A single importer faces no other quantity limit than the whole available quota itself. Moreover, the quotas are not split into quarterly subquotas.

DGAgri quota management is more complicated and, therefore, has often been considered in the literature to be a hindrance to preference utilization (e.g., Bergschmidt and Hartmann 1998, 97–123; Kiss 1994). This justifies the presentation in table 9.1 of the main features of the management and their potential and actual impacts on the quota utilization by Romania. Actual impacts of the various features of the quota management were partly assessed during meetings with agro-processors and/or traders in Romania.

Table 9.1 DGAgri Quota Management and Impact on Quota Utilization

Feature	Potential Impact	Effective Impact
Quarterly subquotas.	Constraint at peak seasons for some products.	Probably low in Romania given the under-utilization of preferences.
Compulsory guarantee deposit by the EU importer for most products, except live bovines and sheep.	Could prevent utilization of preferences in markets where supply is volatile, as in the first years of transition, and therefore import is risky. The guarantee is, however, necessary to ensure that the import takes place.	No difficulty reported by the traders and processors interviewed. Could, however, have had an impact on decisions of EU importers (who could have favored CEECs where the market situation was more stable).
Simultaneous examination, i.e., all requests for preferential import licenses are taken into account (with pro rata reductions if requested quantities exceed the available quota).	Increase in transaction costs compared to simpler systems of quota management such as "first come, first served."	No impact on preference utilization because quotas are not binding for most products (except soft wheat in 1997–1998).
Preferential certificates granted to the EU importer.	EU importers are able to reap the major share of the benefit attached to the preference.	Romanian exporters do not appear to benefit from better prices (interviews).
Overall impact on level of preference utilization.		Probably minor given that the additional transaction costs attached to the quota are far from offsetting the preference margins.

To sum up table 9.1, it does not seem that the management system is responsible for the Romanian difficulties in filling the quota. The main reason for this conclusion is that preferential margins under these quotas are often significant, especially for DGAgri products. This makes their utilization by EU importers highly profitable, even if their complicated management increases transaction costs. The difference between ad valorem equivalents of tariff under MFN and association conditions is often significant; around 100 percent for beef meat, 80 percent for common wheat, and 50 percent for live sheep. However, some features of the preferences might have had adverse effects on preference utilization in the first years of the agreement. For instance, the compulsory deposit of a guarantee could have deterred some EU importers from dealing with Romanian exporters, if the market situation was judged unstable and the risk not to be able to use the preference too high. In the same vein, the rather complicated management of the preferences could have had adverse effects at a time when Romanian traders and processors were not familiar with the agreement itself and with the carrying out of business with the EU. However, no evidence of such adverse impacts in the first years of the agreement was collected during the investigations in Romania.

Unlimited Preferences

Quantitatively unlimited preferences are granted for products that are relatively important for the country. These products, along with the highly processed products under Protocol 3 (the so-called non-Annex II products), accounted for over 31 percent of Romanian agricultural exports to the EU in 1995 and just under 23 percent in 1996. This is clearly not marginal in the export structure of Romania. Annex II of the Treaty of Rome includes non-processed agricultural products and "first-processed" agricultural products (e.g., salami, cheeses). Products which have undergone a higher level of processing were excluded from Annex II and form the so-called non-Annex II products. These include such food products as confectionery, products based on cereals (such as cornflakes), pasta, preparation for soups, and beer.

In general, the duty cuts for these products are limited, as the full tariffs are often low. The maximum preference cut is of 20.4 percent ad valorem (for preparations of fruit), but for 55 percent of the products it is 5 percent or less. With such low levels, most of these concessions can be considered to be of marginal value. It is widely known that the impact of duties on trade is more than proportional to their value. Thus, if the impact of low duties is very limited, hence too are the preferences attached to such duties.

Rigidity of the Preferences

The rigidity of the preferences granted by the EU to the associated countries has been pinpointed in several studies (Duponcel 1998a; Frohberg and Hartmann 1996). This rigidity is even more striking when compared to the high volatility of CEECs' trade during transition (at least at the beginning of the transition period). To allow the EAs to play a more important role in the restructuring of the agri-food sector, all products, both "traditional" ones and new ones emerging as a result of the transition process, should be granted preferential access. Otherwise, the EAs could become a tool for maintaining the export of unprofitable products because they benefit from preferential access at the expense of other products.

The shifts in the export structure of Romania can easily be illustrated by calculation of a distance index *I*.

$$I = \frac{1}{2}\sum_i \left| x_i^m - x_i^n \right|$$

where

x_i^m and x_i^n = the shares of product *i* in total export at years *m* and *n*

The index is zero if the export structures are identical and one if they are completely different, that is, no product in common at the investigated level of trade nomenclature. Calculations were made for 1990, 1993, and 1996 at the CN-6 digit level[2] for several associated countries. In the case of Romania, calculations were also carried out yearly using the base period 1988–1989, chosen as a reference for the preferences (table 9.2).

The results presented in table 9.2 compare Romania with Bulgaria, Hungary, and Poland. They show that the shifts in export structures over time have indeed been significant. This gives strong backing to the proposal to increase the flexibility of the preferences in order to allow EAs to develop along the lines of restructuring CEECs' trade. It is not surprising that the biggest changes in export structures are observed in Romania. Such a significant shift in trade structures is only partially captured by the preferences granted by the EU. At least this evolution is a strong argument for the need to increase the flexibility of the preferences granted to Romania.

Table 9.2 Distance Indexes between Export Structures, 1988–1996

Country	**88–89/90**	**88–89/91**	**88–89/92**	**88–89/93**	**88–89/94**	**88–89/95**	**88–89/96**
Romania	**0.49**	**0.34**	**0.36**	**0.60**	**0.64**	**0.73**	**0.70**
				Index 90/93			Index 90/96
Romania				**0.42**			**0.60**
Bulgaria				0.32			0.49
Hungary				0.24			0.47
Poland				0.32			0.43

Note: as calculations were made with two sets of data, the benchmark for the first row is 1988–1989 and for the other rows it is 1990.

Source: own calculations from Commerce Exterieur (COMEXT) data (index 90/93 and 90/96 from Duponcel 1997)

In order to achieve flexibility, the quotas have to be defined at broader product category levels (e.g., "poultry" instead of a list of CN-8 digit positions). It turns out that the main improvement achieved in this respect for the second period of the EA (mid-1997 to mid-2001) concerns the cheese quota. A significant share of Romanian cheese exports was not covered by the quota up to mid-1997. The list of items included in the quota was ill-defined, disregarding some of the most important export products. A similar problem, which has not been addressed for the second period of the agreements, can be illustrated by the quota of chicken meat; it does not include boneless chicken cuts although they are an important export item.

The Coverage of the Agreement

Few calculations of the coverage of the Romanian agreement are available in the literature. Coverage is defined here as the value of exports of products which are included in the agreement in comparison to the value of all products which are subject to duty payment (full or reduced). Duponcel provided some estimates for Tracy (1994), also published in Duponcel (1996). The coverage was calculated for 1989 and was for 70 percent of the value of Romanian exports (products entering the EU free of duty on MFN basis were excluded from the calculations). Corresponding figures for 1995 and 1996 were 67 percent and 63 percent, respectively. The decreasing trend of the coverage provides evidence that the agreement is unable to accommodate the shifts of the Romanian export structure. It should also be emphasized that, right from the start, the coverage of the agreement was partial. About one third of Romanian exports, subject to tariff barriers, was excluded from the agreement. This is not consistent with the initial arrangement that preferences were to cover traditional trade. It also falls short of the requirement at the WTO that in a free trade agreement elimination of barriers should concern "substantially all the trade" (Article XXIV of GATT).

If the quota levels for mid-2000[3] and the exports of 1996 are taken into consideration, the coverage increases by a mere 4 percent. Binding quotas are a constraint for a limited number of products, some of which are of primary importance for Romania, such as wine. The main constraint is that a number of products are simply not covered by the agreement.

Looking at the coverage according to the main subsectors is insightful. It is particularly low for fresh vegetables and fruit, which have coverage below 50 percent, preparations of vegetables and fruits, and beverages and spirits. These are probably the subsectors where Romanian advantages in export are currently the highest.

Romanian Preferences Compared with Those of Other Associated Countries

The comparison of Romanian preferences with those of other associated countries highlights the disadvantages of the Romanian case. The coverage of the agreement is better for such significant agri-food exporting countries as Bulgaria and Hungary; 94 percent for Bulgaria, 77 percent for Hungary, compared with 70 percent for Romania, (Duponcel 1996).[4] Romania had the worst coverage of the three countries although its exports to the EU were the lowest (ECU 87 million in 1989, against 131 million for Bulgaria and 665 million for Hungary in 1990). As was mentioned, the export structure of Romania changed to a greater extent than that of the other associated countries. Therefore, the coverage deteriorated to the greatest extent.

Furthermore, it turns out that the principle of determining the in-quota quantities based on actual past trade was implemented with variable rigidity. A good example of such discretion is the quota of soft wheat granted to Hungary. Although Hungary exported about 500 tonnes of soft wheat annually during the reference period, it received a quota of 170,000 tonnes, which was increased to 232,000 tonnes for the marketing year 1996–1997. This can be compared with the initial quota of 14,400 tonnes granted to Romania, despite the fact that Romania exported 27,000 tonnes of soft wheat in 1988.

Table 9.3 Shares of Preferences Granted to Romania in Total Preferences Granted to CEEC-6 for the Marketing Year 1994–1995 (tonnes)

	Total Preferences	Romanian Quota	Share of CEEC-6 (%)
Beef and meat preparations	19,192	1,390	7.2
Live pigs and pork products (1994)	77,323	13,317	17.2
Sheep (1994)	29,054	690	2.4
Poultry	67,351	1,022	1.5
Dairy products	18,366	1,333	7.3
Grains	351,926	20,220	5.7

CEEC-6: Bulgaria, the Czech Republic, Hungary, Poland, Romania, and Slovakia
Source: own calculations from Duponcel (1997)

The modest size of tariff quotas granted by the EU to the CEECs has been widely criticized (e.g., Duponcel 1996, 1998c; Tangermann 1993). In this respect, a distinction has to be made between Romania and Bulgaria on the one hand, and the countries of central Europe, which signed EAs earlier, on the other. Preferences granted to Bulgaria and Romania were smaller than those granted to the countries of central Europe. This stems mainly from the fact that past export flows of the two countries were much smaller. However, the EU also adopted a tougher stance with Bulgaria and Romania than with the first associated countries, probably because the economic environment was less favorable than it had been before. Table 9.3 indicates the share of Romania in the main preferences.

Most figures, except those for the pig subsector, display the marginal significance of the Romanian quotas. In most cases, the situation is similar for products administered by the Customs. In the pig subsector, exports of pig meat to the EU stopped in 1993 owing to the new veterinary policy enforced by the EU with regard to swine fever. Vaccination against swine fever is banned in the EU and import of vaccinated animals is therefore not allowed.

Due to the low quota levels, Romania will often be crowded out to the benefit of countries endowed with larger quotas. Romanian quotas will be of interest to the EU importers only if they have not satisfied demand by the utilization of the larger quotas of the other associated countries. This is the case with concentrated apple juice. EU importers prefer to import from Hungary, Poland, or Bulgaria, each of which has a preferential quota exceeding 5,000 tonnes, and only then import from Romania with its 2,000-tonne quota. Under such conditions, Romania is very likely to receive less-favorable prices than its competitors.

The fact that Romania did not obtain satisfactory preferences in agriculture, at least when compared with the other associated CEECs, also reflects the way in which the country negotiated the agreement for both the first and second five-year periods. A number of facts indicate that Romania did not negotiate its agreement properly. At the concluding stage of the negotiations of the agreement, Romania was legitimately dissatisfied with the preferences granted by the EU. Therefore, it attached a unilateral declaration in which it listed the products for which it considered that preferences should be increased. However, this list was far from exhaustive. For instance, it did not include cheese, although the coverage of the agreement was extremely poor (below 25 percent) and the CN-8 positions listed in the agreement were far too restrictive. The EU increased some of the preferences for the second period of the agreement. However, this "improvement" was based on an inadequate monitoring of exports to the EU by the Romanian side. Indeed, additional preferences were granted for products included in the above-mentioned Romanian declaration, but of which very small amounts, at best, were still exported by Romania. At the same time, products more important for Romania, which moreover were not considered as sensitive by the EU, were disregarded. Romania does not appear to have adopted a systematic and consistent strategy for its negotiations. The present monitoring of the bilateral trade flows does not meet the necessary standard. Moreover, Romania's requests for additional preferences are not always in line with the principles of the agreement. For instance, Romania requested preferences for products not exported at the time the request was sent to the EU. Such a request is extremely likely to be turned down by the EC simply because preferences are granted to products actually traded. Of course, this does not concern products for which EU MFN protection can be considered as prohibitive and in which Romania could assume it has a comparative advantage. Possible instances of this are wheat and sheep.

Impact of the Preferences on Agricultural Exports of Romania

Romanian agri-food exports to the EU have increased steadily since 1990. Under the EA, exports increased from ECU 80 million in 1993 to ECU 135 million in 1996. However, it is not possible to assert whether this increase was a consequence of the agreement or rather a sign of the recovery of some Romanian agri-food subsectors after the production slump of 1990–1992.

Exports of fresh vegetables, fresh fruit, live animals, and beverages have increased dynamically under the EA. In other subsectors, exports have either not increased greatly or have even contracted, as in the meat sector. It is possible that the agreement has fostered the exports of fresh vegetables, live animals, and beverages (wine), but its role in these evolutions is not easy to detect. For example, the increase in exports of fresh vegetables is significant, from ECU 10 million in 1993 to 20 million in 1996. However, the coverage of the agreement for fresh vegetables was 43 percent in 1995 and 28 percent in 1996. This suggests that a significant part of the export growth might have been in products that were not covered by the agreement. The role of the wine agreement in the impressive increase in wine exports is probably more important.

Evolution of the Share of the EU in Total Romanian Agri-food Exports

Measuring the impact of the EAs in the countries in transition is a difficult task, since those economies are experiencing unprecedented reforms which have strong effects on production and economic organizations. In the case of agriculture, the most common feature among all CEECs during transition was the fall in output. This output contraction might have directly affected the utilization of some preferences by cutting export supply availability. However, this does not mean that the agreement has not had any impact. It could be argued that, even if the value of exports to the EU has not increased, it is possible that the share of exports to the EU in total Romanian exports should have increased. This should result from better export conditions in the EU market than in other countries' markets, due to the additional competitiveness granted by the preferences. In order to consider the agricultural situation, analysis of the shifts in the EU shares in the total Romanian agricultural exports was carried out. Analysis at a disaggregated level would allow for differentiation between products covered and products excluded from the agreement. However, only aggregate figures were available.

Table 9.4 shows that in fact the EU's market share of Romanian agri-food products has decreased considerably since the beginning of the transition. As the EA was the first of a series of trade arrangements signed by Romania, an increase in this share should have been observed if the additional competitiveness of the Romanian products on the EU market had helped Romania to export to the EU. The poor impact of the agreement is clearly apparent from table 9.4.

Overall, the EA appears to have had very limited impact on the volume of exports. A likely explanation for this result is the low non-price competitiveness—the quality of a number of products being insufficient to comply with the EU market requirements. Under these conditions, export to less-demanding markets is easier.

Table 9.4 Distribution of Romanian Exports according to Destination, 1989–1996 (percentages)

Country	1989	1990	1991	1992	1993	1994	1995	1996
OECD	56.0	53.6	62.6	51.6	49.6	42.6	45.6	40.2
– EU[a]	**48.7**	**45.7**	**51.4**	**42.5**	**33.6**	**32.0**	**30.3**	**21.5**
– OECD-CEECs[b]	0.0	0.0	0.0	0.0	0.0	0.0	0.0	6.1
– North America[c]	1.6	5.1	0.9	0.9	0.8	0.6	0.2	0.3
– Other OECD[d]	5.7	2.9	10.3	8.2	15.2	10.0	15.0	12.4
CEECs[e]	6.6	11.4	5.5	11.8	8.2	12.3	8.1	4.3
CIS	22.9	17.6	9.8	6.2	19.1	19.0	11.8	18.2
Other	14.6	17.3	22.2	30.4	23.0	26.1	34.5	37.3

[a] Austria, Finland, Sweden are included in all years, 1989 to 1996
[b] Czech Republic, Hungary, Poland in 1996
[c] USA and Canada
[d] Including Mexico
[e] Including Albania, Bulgaria, Estonia, Latvia, Lithuania, Romania, Slovakia, Slovenia for all years, and the Czech Republic, Hungary,and Poland 1989–1995
Source: National Commission for Statistics (1998)

Quota Utilization

The level of utilization of the quotas for the CAP products, monitored by DGAgri, is generally low. The highest level was achieved for the sheep quota, which has been completely utilized. The utilization of the beef quota shows a declining trend. Utilization was 72 percent in 1994–1995, but has steadily decreased and fell to 4 percent in 1997–1998. As mentioned, since the beginning of the implementation of the EA, Romania has been unable to use its quota of swine meat for veterinary reasons. The situation has been similar for sausages, but slightly better for preserved meat. The level of utilization of the quota for chicken has not been satisfactory and has never exceeded 40 percent. The utilization of the quota for soft wheat has been unstable. It was satisfactory in 1995–1996, decreased to 65 percent in 1996–1997, and was fully used in 1997–1998. Exports of live bovine animals have been relatively dynamic.

The levels of quota utilization for the remaining products monitored by the EU Customs Administration are better than for the core CAP products, but far from satisfactory. Several quotas (for raspberries, natural honey, nuts, plums, cherries, preserved turkey, apple juice, and wine) have been completely utilized. However, the level of utilization of the other quotas has been either nil or very low. All quotas granted for transformed non-Annex II products have not been utilized at all.

So-called trade outside the quota, which occurs when the actual trade is at MFN tariffs even though the quota has not been fully used, has been observed in several studies (e.g., Duponcel 1997). In the case of Romania, the calculations show that trade outside the quota contracted significantly from 1995 to 1996. This tends to indicate that non- or underutilization of the concerned preferences in 1995 was due mainly to insufficient information being available to traders about the existence of the preferences. This problem was the most acute in the fresh fruit and vegetables subsector.

Impact in Terms of Additional Export Revenue

The sharing of the profits attached to the preference margins between CEEC exporters and EU importers has been a matter of dispute between the associated countries and the EU. The concern is that EU importers capture the main portion of the additional revenue since the EU side manages the preferential quotas. In theory, the ability of the exporters of the associated countries to reap part of the benefit depends mainly on three factors. These are the level of competition in the export supply market, the extent of informational asymmetry faced by the exporters, and the quality of the business relationship established between the exporter and the importer.

In practice, in the case of Romania (and probably also in the other CEECs), some exporters have been able to secure a small part of the profit attached to the preference. This was indicated in the study in interviews and questionnaires with the interested agents. Exporters who have been successful in negotiating on the export price are often those companies with FDI originating from the EU. Clearly, in the remaining cases the balance of power has been unfavorable to the Romanian exporters.

Applying the methodology used in previous work on the Hungarian EA (Duponcel 1998b), the actual benefit gained by the Romanian side in terms of price increases has been estimated. For

1996, this benefit was equivalent to a mere 2.3 percent of the value of agri-food export of Romania to the EU.

Quota Underutilization: Main Reasons Pertaining to the Romanian Side

In this section, the main impediments to export as perceived by the Romanian operators are briefly summarized. The compliance with the EU sanitary and veterinary regulations is discussed in some detail as it has often been asserted that the stringent regulations of the EU are a strong impediment to the utilization of the preferences. A further factor addressed in this section is the role of dissemination of information about the EA.

Overall, the poor results achieved by Romania in utilizing its preferential quotas are explained by reasons pertaining to the current situation and performance of the Romanian agri-food sector. The shortcomings of the preferences might have also contributed to the low level of utilization, for example, too small in-quota quantities for some products, but this is marginal. The preferences could be changed. However, this would not dramatically improve the situation. The main constraints by far are in Romania. The poor results achieved by Romania reflect the slow progress of reforms in agriculture and food processing. More specifically, the present difficulties of the food industry, the restructuring and privatization of which have been long delayed, explain a good deal of the poor trade performance of Romania.

Main Impediments to Export Relating to EU Preference Utilization

Table 9.5 summarizes the main impediments to exports to the EU and utilization of the preferences by Romanian operators.

Table 9.6 presents the main elements regarding sanitary and veterinary requirements. The problems faced by Romanian operators in their exports to the EU were identified during interviews with traders and processors.

It appears that several firms licensed for export to the EU keep their eligibility for certification and are, therefore, inspected every year, although they have not exported to the EU for a few years. These enterprises maintain the license in order to be able to export again to the EU market when circumstances improve. However, maintenance of the sanitary and veterinary standards is quite costly, particularly in plants working well below capacity. Most state-owned slaughterhouses work at about 10–20 percent capacity; the few exporting are those working at full capacity. Most of these enterprises currently face harsh financial difficulties and some of them might go bankrupt. Most private de novo companies do not attain the sanitary and veterinary standards necessary for export to the EU. They do not seem interested in improving their standards or, more exactly, in designing their facilities according to the EU standards, because currently their main target is the domestic market. Therefore, it may well be possible that, within a few years, the number of slaughterhouses and dairies licensed to export to the EU will decrease. As regards dairies, this tendency has already been pronounced. The number of firms licensed to export to the EU was forty in the 1980s and only seven in 1998. For poultry, out of the three licensed firms, only one exports to the EU.

On the whole, the EU requirements for obtaining an export license do not appear to be more stringent than the requirements of other major importers, such as the United States of America. Meat processors and dairies are well aware of the need to achieve high sanitary standards in order to be able to export to demanding markets and, in general, do not question the stringency of the procedure. Obviously, in future, all companies will have to comply with these standards in order to at least be able to stay in the Romanian market. The main problem in the field of sanitary standards is that a significant number of the new private firms, filling the vacuum left by the state sector, are not interested in complying with the EU regulations, since they primarily target the domestic market. Therefore, even if there is currently no shortage of licensed firms for export to the EU, the situation could worsen in the medium term.

Quality of Information Dissemination

Currently, the circulation of information about the EA is not a major constraint on preference utilization. As already mentioned, the current situation of the agri-food sector is the main impediment.

Table 9.5 Main Impediments to EU Exporting and Preference Utilization, Romania

Nature of the Problem	Significance	Description
Cost of credit	High	In the present conditions for credit, agri-food firms are very reluctant to take on new credit. Only firms with foreign capital are able to find credit outside Romania.
Financial capacity of firms	High	Most enterprises have limited financial capacity. Those enterprises, which have succeeded in achieving good export deals, are often the ones with foreign capital.
Cost of raw materials	High	The high cost of raw materials makes competition in export markets extremely difficult.
Quality of raw materials	Medium	In several subsectors: grains (problems at all stages); meat (low genetic quality). The quality of the raw material is considered satisfactory in the integrated complexes.
Production costs	Medium to High	Apart from the cost of raw material, processing costs are high in the existing units (huge overcapacity, high energy costs, etc.).
Quality of exported products	Medium	The quality of Romanian products is not perceived as a constraint apart from in some subsectors such as grains. Marginal exports of bad quality products (such as in the wine sector) should be firmly prevented (by efficient quality controls) as they can endanger the credibility/reputation of whole subsectors and threaten results already achieved.
Competitiveness	High	A number of aspects mentioned in this table contribute to low competitiveness. The competitiveness of some other CEECs is artificially increased by higher support to agriculture (e.g., Hungary with export subsidies).
Infrastructure	Low	The physical and communication infrastructure does not appear to be a significant problem. The situation is however different for grains for which there is an obvious bottleneck at the port of Constanza and with storage facilities/efficiency. The situation will improve with forthcoming additional terminal facilities.

Table 9.5 cont'd

Nature of the Problem	Significance	Description
Transport costs	Low to medium	The EU market is not immediately accessible to Romania.
Unpredictability of government policies	Medium	A stable environment in terms of economic policies and the legal framework is needed for the development of production and trade. It is difficult for enterprises to draw up mid-term plans in a fluctuating and uncertain context. However, the situation has improved since the end of 1996.
Protracted privatization process	High	Restructuring and privatization in the food sector started late and are far from completed.
Low level of FDI	Medium	Owing to the economic context and the political instability, FDI has been slow to come to Romania and is still small. The macroeconomic setting, the reluctance to reform until 1996, and the lack of transparency in the privatization process had a deterrent effect on FDI.
EU sanitary and veterinary regulations	Low	The EU market is not considered by traders/processors to be more difficult to enter than other large markets such as the United States of America. The EU regulations in these matters are not perceived as hidden protectionism.
Non-tariff barriers	Low to medium	Apart from a few subsectors, no problem of non-tariff barriers. The main problems are said to arise for fresh fruit and vegetables (entry prices). No problem with minimum prices for soft fruit.
Market information, information on the EA	Low to medium	For the use of preferences, information appears to be inadequately circulated. However, companies which are most interested in exporting to the EU are able to find information using their own resources. The role of the administration could be improved.

Source: interviews, questionnaire, and literature

Table 9.6 Sanitary and Veterinary Requirements and Difficulties in Romanian Exports to the EU

Subsector	Problems Encountered
Live bovines	– There is a fifteen-day compulsory quarantine period that can impede exports at peak seasons, given the limited capacity of the quarantine facilities. – Identification and certification of animals appear to be satisfactory.
Beef	No particular problem, seven enterprises are licensed to export. Most of them do not currently export to the EU.
Dairy products	EU experts visited Romania in July—August 1997. The verified units had their licenses confirmed subject to requirements of implementing improvements and the provision of clarifications by the Romanian authorities (animal identification and registration, farm recording, milk microbiological quality). Requirements are not considered to be more stringent than for the U.S. market. Up to 1997, nineteen dairies were licensed to export to the EU. In 1998 this number was cut to seven. Most of these dairies do not currently export to the EU.
Pork	Romania systematically vaccinates against classical swine fever. The EU decided to prevent imports of vaccinated animals and products of these animals from 1993. A shift in the Romanian veterinary policy toward the EU would imply a large deconcentration and decentralization of the pig production. Twelve slaughterhouses are licensed to export to the EU.
Prepared pork and sausages	As a result of vaccination against swine fever, only products that have been treated to at least 72°C can be exported to the EU market. Under such conditions, only preserved products can enter the EU. An exception is dried sausages, such as Salami Sibiu, which can also enter the EU market according to the sanitary and veterinary national agency, provided that meat is dried for a period of over 90 days.

Table 9.6 cont'd

Subsector	Problems Encountered
Live sheep and sheep meat	– Romania exports only live sheep. No specialized slaughterhouse is licensed for export to the EU although some projects seem to be under consideration. Some exports can, however, take place through unspecialized slaughterhouses. A sectoral study reckons that no more than three or four lamb slaughterhouses, meeting international requirements, would have to be envisaged for future exports. – For live animals, the EU imposes a thirty-day quarantine, which is quite difficult to implement for lamb since they have to be separated from the mother ewe shortly after their birth (resulting in loss of weight at the quarantine farm). Regarding other export markets, for which Romania has bilateral conventions, the quarantine period is reduced to fifteen days, which is less harmful.
Honey	Since 1998, residues of streptomycine in honey must not exceed 0.2 ppm per kg. The limit was 0.3 before. This limit is likely to be cut further in the next years. There is only one laboratory in Romania for honey where the control can be carried out.
Poultry	No particular problem. Three enterprises are licensed to export, of which only one is actively exporting to the EU.
Wine	Romanian products do not face particular difficulties in entering the EU market. However, some exports fail to meet the minimum quality requirements (exported products do not correspond to the quality ordered, etc.). The quality control of exports has to be improved and enforced more effectively by the Romanian authorities in order to avoid this kind of problem which endangers the reputation of the whole subsector.

Source: interviews

However, interviews with traders and processors have shown that the circulation of information could be improved. Although it would not dramatically change the whole picture, it might be useful at least in particular sectors.

Besides the publications in the *Official Journal of Romania*, information on the EA is published in several business reviews which, however, are read by a minority of operators in the food industry. Staff at the MoAF and at the Ministry of Industry and Trade (MIT) can also be contacted with questions related to the EA.

The analysis indicated that the quality of information reaching the operators could be improved. Often traders and processors do not have up-to-date information on the level of utilization of preferential quotas. A minority of operators even seems to ignore the EA. Interviews indicate that those companies that are well informed are also the ones with good connections with the EU and, quite often, have western European partners, joint ventures, or other types of FDI.

The current dissemination of information on the agreement could be improved if it were channeled through specialized agri-food publications. One option is to inform the producers, processors, and/or traders' associations concerned on a regular basis, and they would then use their own information channels to reach their members.

The monitoring of the utilization of the preferences is made about every quarter at MoAF on the basis of the information received either from the Romanian delegation at the EU in Brussels or directly from EU Customs for products administered by them. However, the results of this monitoring are not made available to the operators in a systematic way. For instance, although the wine quota was already binding (the wine quota was 168,880 hl in 1998), no information was provided to the exporters. Their trade partners in the EU were their only source of information.

Systematic monitoring of at least the binding quotas is necessary (e.g., at peak seasons) in order to facilitate price negotiations with EU importers. This would decrease the informational asymmetry, which currently significantly affects the Romanian exporters, and assist them in their search for trade partners.

This monitoring could be carried out by the administration (MoAF or MIT) and systematically forwarded to the operators. It could also be carried out directly by the operators for the products they are interested in. However, they should be informed about the means available to them for performing the monitoring. For products administered by the EU Customs, it could be carried out directly with information obtained from the Internet. For DGAgri products, operators would have to rely on information provided by the Romanian administration.

Conclusion

The overall impact of the EA on the Romanian agricultural trade with the EU is insignificant for most products. However, the wine agreement contributed to the development of exports of wine.

There are several explanations for the poor results from the EA. The current difficulties faced by the Romanian agri-food sector appear to be the main constraint. Under the present circumstances, Romanian agri-food products have limited competitiveness on the EU market. The very features of the management of the preferences by the EU are not responsible for the difficulties faced by Romania. They may have had adverse effects in the first years of the agreement when the situation of Romanian agriculture was less stable; however, no such evidence was reported. Similarly, it cannot be said that the veterinary and sanitary controls imposed by the EU are responsible for the difficulties of Romania.

The coverage of the agreement is very restrictive. First, this reflects the fact that the negotiations of the EA took place at a time when the EU's willingness to open up its market had diminished. Second, this is the result, to some extent, of the poor handling of the negotiations by Romania. In this respect nothing else explains the striking difference of coverage between the agreements of Romania and Bulgaria.

The EU granted additional or enlarged preferences for a number of products for the second period of the agreement, from mid-1997 to mid-2001. However, this did not significantly change the global picture.

Since the study was completed, further trade liberalization between the EU and the ten associated countries has taken place under the framework of the pre-accession process as an initiative of the EU agreed in spring 1999. The basis of negotiation between the EC and the ten associated countries was the so-called double-zero option: no import duties and no export subsidies for trade within stipulated annual quotas in both trade directions. Negotiations were concluded in the first half of 2000 and the new preferences have been implemented since July of that year. In the case of Romania it can be said that the final outcome falls short of the initial aims. Concerning the most important quotas for Romania, the EU has granted duty-free access only to a few products, mainly bovine meat, cheese, and common wheat (see Council Regulation 2435/2000). For other important products (e.g., fruit and vegetables) the level of the preferential duty has not been changed. In addition, only the size of the sheep quota has been significantly increased (from less than 3,000 tonnes in the former protocols to 7,000 tonnes in the new one), while most of the other quotas have remained at their previous levels. With the new provisions, the preferential access to the EU market of Romania has only marginally improved. Most of the shortcomings of the EA described in the present analysis are therefore still valid.

The EA has been considered by the EU as an important instrument for the preparation of Romania for accession. However, as far as agricultural trade is concerned, there is a huge gap between the declarations and the evidence. The agreement has been far too restrictive to foster agri-food exports of Romania to the EU market and to facilitate the process of economic integration. It was hoped that Romania would gain a better access to the EU market with the last negotiations. This opportunity was apparently missed.

Notes

1. This chapter summarizes the main findings of the report written for the project at a time when the author was not a staff member of FAO. Views expressed in this chapter are the author's only and do not necessarily reflect those of FAO.

2. CN refers to the Combined Nomenclature used by the EU. Its most disaggregated level is with eight digits. The most aggregated level is two digits (chapters of the nomenclature).

3. For wine, it is assumed that the quota for 1997 (158,880 hl) is given an annual 5 percent increase until mid-2000. In this calculation, the additional preferences granted by the EU (all positions for cheese, new positions for fruit juices, etc.) are also taken into account.

4. Based on trade figures for 1990 for Bulgaria and Hungary and 1989 for Romania.

References

Bergschmidt, A. and Hartmann, M. 1998. Agricultural trade policies and trade relations in transition economies. *Agricultural Policies in the Baltic Countries: Proceedings of the Parnu Seminar held in 1997*, 97–123. Paris: Organization for Economic Cooperation and Development.

Duponcel, M. 1996. Accord d'Association Hongrie-UE, impact de l'ouverture du marché communautaire sur l'exportation agro-alimentaire hongroise. INRA-ESR, Série *Etudes et Recherches*, 104.

———. 1997. *Europe Agreements between the EU and the CEECs: Functioning of the Trade Preferences Granted by the EU in Agriculture, Possibilities of Improvement and assessment of quota utilization.* Report to Unit VI-A-1 of General directorate for Agriculture. European Commission.

———. 1998a. *Impacts of the Europe Agreement on Romanian Agriculture.* Report for the project EU PHARE RO9505-04-01/04 "EU alignment and policy reform/information and debate".

———. 1998b. The Europe Agreement between Hungary and the EU, impacts on Hungarian agriculture. *Russian and East European Finance and Trade* 34:79–96.

———. 1998c. The poor impact of EU liberalisation of agricultural imports from Central and Eastern Europe, failure of the preferences or failure of the associated countries? *Food Policy*, 23:187–198.

Frohberg, K. and Hartmann, M. 1996. *Promoting CEECs' Agricultural Exports through Association Agreements with the EU, Why it does not Work.* Paper presented at the VIIIth congress of the EAAE, Edinburgh.

Kiss, J. 1994. Hungary. In *East-West European Agricultural Trade, the Impact of the Association Agreements.* M. Tracy, ed. Genappe, Belgium: Agricultural Policy Studies (APS).

National Commission for Statistics. 1998. *Anuarul statistic al României.* Bucharest, Romania: National Commission for Statistics.

Tangermann, S. 1993. Some economic effects of the EC agricultural trade preferences for Central Europe. *Journal of Economic Integration* 8:(2) 152–174.

Tracy, M. ed. 1994. *East-West European Agricultural Trade, the Impact of the Association Agreements.* Genappe, Belgium: Agricultural policy studies (APS).

Chapter 10

Rural Development and Pre-Accession Strategy for Romanian Agriculture

Graham Dalton, Bill Slee, and Kenneth J. Thomson

Much empirical evidence exists about the poor conditions of life for an increasing number of rural people in Romania. These well-documented sources (e.g., Chirca and Tesliuc 1999) indicate that the pathway of transition to a market economy has produced a significant increase in poverty in rural areas. Within the existing EU, there is evidence that transfers under the Agricultural, Structural, and Cohesion Funds have not always fulfilled expectations of their contributions to the growth and development of rural economies, mainly through supporting "unproductive" public investments (Efstratoglou et al. 1998; Hervé 1999). It would be a tragedy if this were to be the outcome in the applicant countries where socio-economic problems in rural areas are even more obvious.

First, some terminology requires clarification. Rural development may be understood as the general restructuring and improvement of economic and social activities in rural (nonurban) areas. This interpretation concerns the overall balance of economic activities, for example, between agriculture, forestry, and other land uses, and between land and non-land sectors such as food processing, transport, retailing, and tourism, along with the relationship of social to economic factors. Some would add environmental considerations into this analysis, especially if policy intervention (see below) were significant.

However, in European Commission usage, "rural development" sometimes takes on a much narrower meaning. Here, a strong orientation toward farming and farm-based activities is noticeable, and especially toward "agricultural development" in terms of farm restructuring, investment, and reskilling, and toward "agri-environmental" issues such as soil and water conditions, wildlife habitats, and landscape preservation. This approach attempts to couple the preservation of a substantial structure of "family farms," run essentially on traditional lines, with efforts to deal with the growing conflict between modern farming methods and increasing environmental concerns, whether expressed by visitors, incomer residents, or citizens generally. The wider aspects of rural economic development, such as housing, social services, and the prosperity of businesses unrelated to the food and fiber chain, are generally regarded as being the concern of others. Clearly, this narrower approach derives from the historical development and responsibilities of agricultural policy institutions such as national ministries and the EU Directorate-General "for Agriculture and Rural Development." Despite innovations such as the "integration" of the EU Structural Funds in the 1980s, these institutions have historically been tasked with addressing the conventional "farm income problem" via methods ranging from market support, direct payments, or (as in Romania) more or less direct state regulation and control.

Against this background, "rural development policy" must be approached with some care as to whether what is meant is the complete spectrum of government responsibilities outside the towns and cities, or the more specific policy objectives of agricultural measures. The potential confusion is greater when, as in current-day Romania, one is dealing with a country where most state institutions have been weakened, but are now being asked to address a widening range of "new" issues, including market orientation, infrastructure renewal, and social problems. Moreover, the concern of this book is with eventual accession by Romania to the EU, which, as suggested above, has its own approach to rural development, as expressed through the relevant parts of the acquis communautaire. Thus, this chapter first considers the background to rural development in Romania, and then presents some evidence from recent household/business surveys carried out in rural Romania. This is followed by some commentary on the recently approved Special Accession Programme for Agriculture and Rural Development (SAPARD) plan, and some conclusions are drawn on some elements of a solution to rural development problems in Romania.

Rural Romania

The rural population of Romania accounts for 44 percent of the total population (World Bank 2001). In some counties, there is potential growth in the size of the rural labor force, although overall the rural population of 10.8 million in 1989 decreased to 10.2 million in 1999 (NCS 2000). Levels of human capital, in terms of education and skills, are dramatically lower in rural than those in urban areas. Health and other social services are often inadequate, and much infrastructure, including roads, telecommunications, water and waste management, is in need of upgrading. There is very little support for business in the form of readily available credit, information, and mutual organization, and remarkably little trust among rural people in the capacity of government or other institutions to improve the conditions of living for rural people. All of these facts contribute to a sense of alienation and a lack of rural entrepreneurship and confidence, which are essential to successful rural development.

Rural areas in Romania (as in other CEECs) are characterized by a very heavy dependence on farming, which, in the absence of alternative opportunities for work and difficulties of marketing, has taken on a largely subsistence character. In terms of unemployment, low incomes, and lack of economic opportunities, these rural areas face many disadvantages. Agricultural labor productivity is low, both with respect to the rest of the Romanian economy, and internationally, even after adjustments are made for the part-time nature of much of this type of employment (occupation). Farmers have low levels of skill and knowledge, and as a result do not make effective use of available technology. As explained in chapters 4 and 5, many do not or cannot access credit through banks, and again this leads to low levels of use of simple technologies such as improved seeds or fertilizers. For the most part, the size of individual farms is so small that it is impossible to earn a reasonable level of income from such businesses. Also, although there are signs of improvement in some situations, there are many examples of institutional and market failures in the sector, including delayed payments and high transaction costs within the food chain and, as shown in chapter 8, input prices that are above international levels. The capital stock of farms, especially of machinery, is run down and inadequate provision has been made for retirement.

In spite of its backward character, there are reasons to assert that Romanian peasant farmers may be "efficient but poor" (Schultz 1964). Inefficient markets for both inputs and outputs, the weakness of rural financial systems, and the absence of off-farm work opportunities all combine to create a situation in which the production of a diverse range of subsistence crops and livestock, with scope for the marketing of surpluses, may be an entirely rational course of action in the current conditions. The key to changing this situation lies largely in the creation of an efficient market system (OECD 2000).

There are substantial differences in living conditions and economic opportunity between different parts of Romania. Generally, the extent of poverty is far deeper in the east and south of the country along the Danube plain and in the hilly regions of Moldova. These are predominantly areas of arable agriculture, with a relatively small amount of land per capita, and are poorly endowed with human capital (a high percentage of people with only primary school education). There are also pockets of poverty in some of the mountain areas, but in general, on a range of criteria, the counties and communes of the Carpathian arc outperform those of the plains and the Moldovan border region (Chirca and Tesliuc 1999).

Paradoxically, it is the more mountainous regions, which are agriculturally relatively disadvantaged, that have shown the greatest adaptive capability in the movement toward a market economy. A likely reason for the higher adaptive capability is the fact that collectivization had a less comprehensive and durable impact in the mountainous areas, so that in these areas a more individualistic culture, better adapted to a market economy, has prevailed. In addition, throughout Romania, those farmers who are located closer to market towns have been better able to commercialize their farming operations through the use of local markets.

Recent studies of economies in transition (Davis and Pearce 2000) have indicated the importance of the non-farm rural economy, and they and other authors have suggested that up to 50 percent of rural incomes may be derived from nonfarm sources. This nonfarm rural economic activity is often less well understood than the farm economy, but may provide the underpinning of the new rural economy. In economies in transition where there have been major changes and ruptures, the case for livelihood diversification associated with subsistence farming is strong.

Agricultural Development

As part of a study on alignment of Romania to the EU (Popov, Lubieniechi, and Slee 2001), a household questionnaire survey was carried out in selected communes from the following counties: Albesti in Vaslui county; Corbi in Arges county; and Valea Macrisului in Ialomiţa county. These counties and communes were selected to represent rural communities of various types: Albesti is in a hilly area near the Moldovian border, with a very high poverty index; Corbi is in the Transylvanian Alps; and Valea Macrisului is on a rural plain north-east of Bucharest. In each commune, fifty farm households, representing between 6 and 9 percent of the total, were randomly selected from residential lists. Background questions confirmed the expected predominance of small household sizes, high average ages, and high unemployment.

Table 10.1 shows the range of main sources of survey household incomes. The main income source in Albesti for 44 percent of the survey households was the pension; however, agricultural output and salaries were also considered important. In contrast, salary and agricultural output were reported as the main income source for almost a third of the households in Corbi. Agricultural output in Valea Macrisului represented the main source of income for half of the interviewed households. More generally, although not shown in the table as a main income source, children's allowances (but not unemployment allowances, which were almost negligible) were important to many families, and many households were pluri-active, with a number of subsidiary income sources.

Table 10.1 Main Income Sources (percentage of survey households)

Source	Albesti	Corbi	Valea Macrisului
Salary	20	32	22
Pension	44	18	22
Unemployment allowance	0	0	2
Social help	2	0	0
Agricultural output	22	30	50
Non-agricultural output	0	0	2
Services provided	4	8	2
Casual work	8	8	0
Other	0	2	0
Total	**100**	**100**	**100**

The main constraints on operating and developing farm businesses in the survey areas are shown in table 10.2. The main problems and complaints were either financial, such as low net output prices and high input costs (taxation and lack of credit were not often mentioned), or related to inadequate physical resources such as land and machinery. "Soft" resources, such as market access, labor quality, and advice availability, were less often cited, although allowance must be made for the subjective and perhaps limited experience of the interviewees.

Table 10.2 Main Farm Business Constraints and Problems (percentage of survey households)

Constraints/Problems	Albesti	Corbi	Valea Macrisului
Low prices	100	**91**	**97**
High transportation costs	**67**	**54**	**46**
Small batch size	**67**	**57**	**59**
Late payment	24	20	15
Inadequate machinery	**72**	**36**	**46**
Land area too small	**26**	**66**	**52**
Poor input quality	**46**	**32**	28
Lack of technical advice	6	4	4
Lack of marketing advice	2	6	2
Costs too high	12	8	**32**

The preferred methods of business assistance for the households interested in starting their own business are presented in table 10.3. In all three communes, one preferred method seems to be "visits to other businesses," showing that many want to learn from other people's experience, and perhaps they mistrust "experts" and organizations. However, receiving advice from a consultant was favored in Valea Macrisului.

Table 10.3 Preferred Methods of Business Development Assistance (percentage of survey households interested in starting their own business)

Methods of Assistance	Albesti	Corbi	Valea Macrisului
One-to-one advice from consultant	35	42	**60**
Locally held training course	35	16	15
Expert report	24	21	25
Visits to other businesses	**53**	**63**	**55**
Farmer organization	12	16	20
Producer organization	29	21	15

In order to investigate willingness and ability to take up funding likely to be made available from the EU within the SAPARD program (see below), households interested in obtaining SAPARD project funds—about 40–50 percent of the total—were asked if they would be able to cofinance such a project. In Albesti, 42 percent of the respondents interested in SAPARD funds were able to cofinance 50 percent of the cost of the project, as were 58 percent in Corbi and Valea Macrisului (table 10.4). Many of these intended to use family or own savings, although, especially in Corbi, the importance accorded to long-term commercial loans was significant. However, in two of the three communes, a significant proportion of such persons deemed themselves unable to cofinance.

Table 10.4 Households Able to Cofinance the Project (percentage of households interested in obtaining SAPARD funds)

Cofinancing	Albesti	Corbi	Valea Macrisului
0%	**42**	**0**	**19**
10%	0	5	0
15%	4	0	0
20%	0	5	4
25%	8	21	12
30%	0	11	8
35%	4	0	0
50%	**42**	**58**	**58**

Non-agricultural Rural Development

Alongside the household survey in the EU alignment study, a parallel survey of local entrepreneurs and managers in each commune was carried out among seven to ten enterprises started since 1990. The types of businesses found varied from farm- and (in Corbi) forest-related enterprises to trading activities. Managers had a varied educational and working background and an average age of about forty-five in each commune.

Some of the main problems faced by these managers were the same as those mentioned by the farming households cited earlier: on the sales side, low output prices (especially after transport costs were paid) and late payment; and on the input side, the poor availability and high cost of finance, poor roads, and weak local product demand. The types of assistance that these managers wanted for future business development are shown in table 10.5. Infrastructure was seen as the major development constraint: all of the businesses in Albesti that wanted development assistance, and 80 percent of those in Corbi, identified "direct state investments in infrastructure" as necessary. Many respondents saw "reduced local and national taxes" as desirable, but, more encouragingly, many wanted to learn how to prepare a business plan.

Overall, therefore, it seems that non-agricultural rural development is more highly dependent on credit facilities, and more desirous of business and marketing advice, than as expressed by farmers. Many respondents would like to receive assistance, but identified it as being unavailable and/or did not know how to seek it out.

Table 10.5 Percentages of Businesses Wishing Assistance, by Type of Business Assistance

Type of Business Assistance	Percentage of Businesses Which Would Like to Receive Assistance in the Future for Business Development			Percentage of Businesses Which Would Like to Receive Assistance but Identified It as Being Unavailable		
	Albesti	Corbi	Valea Macrisului	Albesti	Corbi	Valea Macrisului
Prepare business plan	**43**	**70**	**50**	**71**	**80**	**60**
Marketing advice/strategy	29	40	20	57	40	30
Building/architecture services	0	0	0	29	10	20
Prepare loan application	14	20	10	0	10	20
Technical/production advice	0	0	10	14	20	20
Management advice	0	10	20	43	20	30
Direct state investment in infrastructure	**100**	**80**	50	**100**	**90**	50
Reduced local and national taxes	**86**	**70**	**60**	**86**	**70**	**60**
Assist business/ train staff	14	20	10	29	30	20

SAPARD

SAPARD in all the applicant countries was first mentioned in the accession strategy paper of the Madrid Council of November 1995. At the 1999 Berlin Council, a fund of Euro 520 million per year was agreed for the period 2000–2006. The annual SAPARD allocation for Romania amounts to Euro 153 million in 2000 prices (SAPARD Programmes, Romania). SAPARD is important in both a symbolic and a practical sense. Firstly, its launch was the first occasion when the EU demonstrated its enlargement commitment in financial terms to the agricultural and rural communities of the CEECs. Secondly, in implementing SAPARD, a key objective is to familiarize these countries with the administrative systems required for the expenditure of EU funds for development purposes. The proposed EU priorities as set out in the SAPARD regulation are (European Commission 2002):

- to help solve the priority and specific problems in agriculture and rural development (such as improving market efficiency, and adopting EU levels of quality and health standards); and
- to contribute to the implementation of the acquis communautaire (the whole body of Community legislation) concerning the CAP and other agricultural priorities.

These steps are necessary in order to enable accession countries to take part in the EU single market and to act on behalf of the union at external borders.

A candidate country may choose among fifteen specified measures (see table 10.6), according to the priority areas of its Rural Development Program. Of these, nine are predominantly agricultural in nature. Moreover, many involve considerable expense, especially in the area of food processing, so that, with the need to provide national cofinancing alongside EU funds, the choices for using the funds are not easy to make. Romania is no exception to this tendency. Within the defined priority to develop the rural economy, the Romanian plan directs most of the investment to agriculture and its physical processing and marketing facilities. Such a sectoral approach assumes either the existence of adequate supporting services in other areas (and farm supplies of adequate quantity and quality), or that a parallel response to other initiatives, private and public, will be forthcoming in these areas.

Table 10.6 Available SAPARD Measures

Investment in agricultural holdings
Improving the processing and marketing of agricultural and fisheries products
Improving the structures for animal and plant health controls, for the quality of foodstuffs, and for consumer protection
Agricultural production methods designed to protect the environment and maintain the countryside
Development and diversification of economic activities, providing for multiple activities and alternative income
Setting up farm relief and farm management services
Setting up producer groups
Renovation and development of villages and the protection and conservation of the rural heritage
Land improvement and reparceling
Establishment and updating of land registers
Improvement of vocational training
Development and improvement of rural infrastructure
Agricultural water resources management
Forestry, including afforestation of agricultural areas, investments in forest holdings owned by private forest owners, and processing and the marketing of forestry products
Technical assistance for the measures covered by Regulation 1268/1999, including studies to assist with the preparation and monitoring of the program, information and publicity campaigns

Source: European Commission (2002)

Preparation Processes

Preparations for EU-funded rural development initiatives in Romania are a new field of activity, and one putting enormous strain on a few dedicated people within ministries. A plan has to be submitted to the commission before implementation begins. This plan has to be based on sound evidence and be evaluated ex ante, show that consultations with intended stakeholders have taken place, and provide a coherent allocation of resources between measures, by year and possibly by targeted locations over a seven-year period. EU funding is accession driven, and thus SAPARD funding has to be consistent with accession partnerships and the National Program for the Adoption of the Acquis (NPAA). The requirement that SAPARD implementation procedures should conform to EU practice and stringent audit procedures has resulted in much delay to the detriment of intended beneficiaries. By the end of December 2001, only five out of the ten CEECs eligible for SAPARD were cleared for receiving and spending money: Bulgaria, Estonia, Latvia, Lithuania and Slovenia (*Agra Europe* 2001). Romania still does not have an official accreditation of an EU-approved paying agency for SAPARD aid.

A sensible aspiration is that rural actions should be problem driven. Within the EU, great emphasis and pressure is being given to a regional approach with decentralized decisions, a move that in theory should help in keeping assistance closer to reality. In Romania, this intention has had initially an unfortunate side effect, in that the role of the MoAF in rural development was challenged to such an extent that it delayed the start of SAPARD preparations while the new regional organization was set up. It is thus encouraging to see the emergence of informal partnerships or networks of interested ministries and the beginnings of joint actions with regional governments in the planning process. This has been incorporated in the agreed Special Preparatory Program (SPP) financed under the PHARE program (see below). It is also encouraging to see working relationships develop between the EC staff responsible and those involved in pre-accession measures. Perhaps most encouraging of all is the new mood of cooperation between the EU and the World Bank, who, along with private nongovernmental organizations (Willis and Bryden 2000), face a common challenge in EU applicant countries, which have much to learn from each other.

The Special Preparatory Program for Pre-accession Funding

Under the PHARE Programme RO9807-01 "Regional Policy and Cohesion," the SPP aims at preparing Romania for the management of EU pre-accession SAPARD funds. As such, this program is an illustration of revealed preferences for agricultural and rural development in Romania. Under the SPP, Euro 2.5 million, to which a national budget contribution of 25 percent of the value of the PHARE investment (i.e., Euro 625,000) will be added, are dedicated for investment projects of the same kind as those that will be funded under the SAPARD program. A fairly centralized system of approval has been proposed, albeit with representation from a range of ministries. Use will be made of regional offices and a variety of official and nonofficial bodies in the preparation of applications. Priority will be given to projects having a total value of Euro 100,000 or more, including public and private contributions. Where a project supports private sector development, the contribution of the private partners should not be less than 50 percent of the total eligible costs.

Initially, two microregions consisting of twenty "communes" from the Satu Mare county and fifteen "communes" from Calarasi county were targeted for SPP. Subsequently, the SPP monitoring committee decided to extend the eligibility to all Romanian territory. However, the SAPARD implementation mechanism will be tested primarily in the two original microregions, for which a pilot rural development plan, based on SAPARD principles and methodology, has been prepared. Four measures have been selected from the eligibility list (see table 10.6):

- Investments in agricultural holdings (measure 1)
- Improvement of processing and marketing of agricultural products (measure 2)
- Development and diversification of economic activities providing for multiple activities and alternative income (measure 3)
- Development and improvement of rural infrastructure (measure 4)

Support will be available for a wide range of farm investments, including machinery, equipment, livestock improvement, assistance for alternative enterprises in rural areas including tourism, roads, drains, waste disposal, and improved water supplies. Despite its limited scale, the pilot plan illustrates the interdependence of both the main measures and the various submeasures: for example, the dependence of farm and food processing on adequate infrastructure; and the need to find new income-earning opportunities in rural areas, especially in those modernizing agriculture.

Quality and health standards are not only important from an accession perspective but also for Romanian consumers. However, standards vary according to the market segment. As underlined in chapter 9, smaller de novo food businesses may not find it commercially attractive to make the necessary large investments, especially if they serve only domestic markets. Although there are derogations for small plants within the EU, there is a case for public funding for plants to adopt the EU standards and for the industry to adopt health controls as a necessary condition for exporting both to the EU and the rest of the world. In view of the amount of funds involved, a bidding system may be the best way of implementing such measures. From earlier EU experience, a real danger of public funds being used in this sector is the creation of overcapacity, especially if plants which do not meet standards (and usually have lower short-run costs) are allowed to continue in production.

Rural job creation is well recognized as a necessary part of the solution to the farm income problem and also as a rational response to the labor-shedding consequences of modernization in farming and food processing. It is thus a necessary component of a coherent agricultural and rural program. However, the measures in the EC Regulation are not obviously targeted at this priority. Investments and improvements that increase market efficiency will have strong multiplier effects within a rural economy. But the specific measures which might foster job creation and which should be consistent with the relevant acquis are constrained to agricultural situations. Incentives such as business start-up assistance for diversification of economic activities should be available for a range of rural businesses (usually small and medium-sized enterprises), from other funding sources if not from SAPARD. Likewise, vocational training improvements should be utilized to assist those within agriculture to shift to other jobs or, in the case of children from farms, to have sufficient skills to enter nonfarm sectors. The concept of infrastructure should be broad enough to allow contributions from SAPARD to investments in human capital, including health and safety.

In the longer run, it is inconceivable that the present farm workforce in Romania can be sustained in farming. Admittedly, the natural diminution of an elderly farm population may reduce the number of farms, but the average farm size will remain very small for the foreseeable future. It is essential that the economic base of rural areas is broadened. Given the limited local demand (a result of many poor households), it is desirable that the new "motors" of the rural economy can be competitive not just in Romania but in wider European markets. The further development of non-farm rural businesses will thus provide the foundation for future rural prosperity, although farms operated on a part-time basis will continue to satisfy basic subsistence needs.

Over and above the need to consider the development of human capital to support new entrepreneurs in rural areas, there is a need to consider the development of human capital in the supporting institutions, from advisory service to local councils. While many rural people continue not to trust state institutions, it is hard for new partnerships and platforms for development to emerge. The term "social capital" is often used to describe this necessary trust between different individuals and between individuals and the various state bodies, the absence of which causes a major impediment to development.

Conclusion: Comparing Development Prerequisites and SAPARD/SPP Measures

A comparison of the necessary conditions for agricultural and rural growth in Romania with the priorities and measures of SAPARD and the SPP leads to the following conclusions. First, the SAPARD list, in whole or part, is not comprehensive enough to produce rural and agricultural growth and development. A much wider set of conditions needs to be present, including institutional reform, incentives for change, and stimulants to innovation. This is especially the case with rural job creation if the measures are essentially farm based, and if training and education are not geared to fit people to obtain and develop nonagricultural, income-earning opportunities. In consequence, the relevance and quality of SAPARD plans can only be judged in a strategic sense by the way that they fit into a more broadly based rural set of actions. A National Rural Strategy is thus more important than SAPARD plans.

Second, it is not sensible or politically feasible to pursue market efficiency without also considering the distributional impact or fairness of the chosen measures (the same applies to environmental concerns). It will, thus, be necessary to target specific groups, such as individual farms within a given size range, processors who are either exporting or intend to do so, producer associations and groups, women in the rural economy, and regions where migration, deprivation, or environmental loss could occur. One criterion for the prioritization of SAPARD agricultural development measures would be farm business size defined as some cutoff value of annual sales. Plans might reasonably be judged on the basis used for choosing eligible beneficiaries. Another criterion might be the need to avoid widespread failures in farm businesses which have been encouraged to expand into a potentially rewarding sector (here, Romania's fundamental competitiveness needs to be assessed, commodity by commodity, including processing stages) but which lack suitable outlets.

Third, the SAPARD priorities are highly complementary (coherent), in that gains in market efficiency will lead to a need for shifts in occupational patterns, normally out of (or at least not into) agriculture. The quality and food safety priorities will facilitate exports and the retention of domestic markets against imports, and so sustain or improve rural employment. As regards price reporting and quality standards, the acquis will assist in the development and operation of markets. The test of coherence should demonstrate how the various chosen measures fit together and contribute as a whole to rural development.

A final point is to reiterate the importance of flexibility and simplicity. It seems absurd to allocate funds in a precise way for fourteen measures across seven years and possibly by region as well. Fortunately, the emphasis on evaluation in the regulation should allow institutions to learn as experience is gained. SAPARD is not an exercise in central planning, but should be more of an indicative statement of intent based on good analysis and clear objectives. It should also be acceptable to introduce, at a later stage, measures that would not require major funding, but which might in some situations make all the difference.

The choice of agricultural versus non-agricultural rural initiatives will influence overall results. It is very clear from the statistics of labor use in Romanian agriculture that efficiency improvements will involve a shift in labor resources out of the sector. But the creation of additional employment depends on entrepreneurship in the wider rural economy as well as a profitable and growing agricultural sector. Clearly, there is a need for rural entrepreneurs to create products that can be exported out of the rural areas where demand is very weak and the market economy underdeveloped. Given the current weaknesses of the market system, local businesses tend to resort to local markets, even though prices are very low. It is desirable to break out of this cycle. Export earnings either through "invisibles," such as tourism, or products in which rural Romania can be competitive internationally provide the foundation stones for the new rural economy.

Returns are also likely to be higher from actions which use taxpayers' funds for public purposes, such as education, training, health, the environment, infrastructure, and other services which the market would not normally provide at the appropriate level. In Romania's case, measures to improve or even establish market mechanisms would be included here, such as information provision. Targeting is also essential given the small amount of funds available. The suggestion made by Hazell (1998) that support for small family farms can be supportive of both economic and social objectives may, if true, be a fortunate finding for Romania. Finally, SAPARD is encouraging the setting up of organizational arrangements for development. Two conditions which, if fulfilled, are likely to improve success are decision-making systems which involve the beneficiaries, and administrative support which takes on the role of "animation" rather than passive bureaucracy.

Acknowledgments

Much of the fieldwork on which this chapter is based was carried out by Simona Popov and Simona Lubieniechi while attached to an EU-funded pre-accession project under Dr. Simon Davies. Their commitment and hard work were much appreciated.

References

Agra Europe. 2001. No. 1982. December 14.

Chirca, C. and Tesliuc, E-D. 1999. *From Rural Poverty to Rural Development*. Bucharest, Romania: World Bank/National Commission for Statistics.

Davis, J. and Pearce, D. 2000. *The Non-Farm Rural Economy in Eastern and Central Europe: Characteristics, Importance and Policies.*, ODI: London. Unpublished report.

Efstratoglou S., Psaltopoulos D., Thomson K. J., Snowdon P., Daouli J., Skuras D., Kola J., and Nokkala M. 1998. *Structural Policy Effects in Remote Rural Areas Lagging behind in Development.* Final report to European Commission on STREFF project FAIR3 CT96-1554.

European Commission. 2002. *The Enlargement Process and the Three Pre-Accession Instruments: PHARE, ISPA, SAPARD.* Brussels, Belgium: Enlargement Directorate General.

Hazell, P. 1998. Agricultural Growth, Poverty, and the Environment: an Introduction. *Agricultural Economics* 19:ix–xii.

Hervé, Y. 1999. *EU Regional Transfers and Macroeconomic Fiscal Reactions in the Cohesion Countries: An Empirical Analysis.* University of Saarland, Germany. Report presented to the World Bank, Washington, March 9 1998.

National Commission of Statistics (NCS). 2000. *Annuarul Statistic al Romanei, serii de timp 1990–1999.* Bucharest, Romania: National Commission of Statistics.

Organization for Economic Cooperation and Development. 2000. *Romania: Country Study.* Paris: Organization for Economic Cooperation and Development.

Popov S., Lubieniechi, S., and Slee R. W. 2001. *Opportunities and Constraints for Rural Development in Romania.* PHARE Project RO9804.03.01 EU Alignment and Policy Reform.

SAPARD Programmes 2000–2006. Romania. http://www.europa.eu.int/comm/agriculture/external/enlarge/countries/romania/overview/rom_en.pdf

Schultz, T. 1964. *Transforming Traditional Agriculture.* New Haven, Conn.: Yale University Press.

Willis, P. and Bryden, J. 2000. *The Implementation of Agenda 2000 in Rural Areas of Eastern and Western Europe.* Report of the 1999 Arkleton Trust Seminar held 19–21 March. The Arkleton Trust: Enstone, Oxon, UK

World Bank. 2001. *World Development Indicators 2001.* Washington, D.C.: World Bank.

Chapter 11

Modeling the Economic Costs and Benefits of EU Accession

Kenneth J. Thomson

The main objective of the study[1] described in this chapter was to identify and estimate the economic welfare effects of future Romanian accession to the EU, and, in particular, the adoption by Romania of the EU's CAP. It is assumed that the date of formal EU accession and CAP adoption by Romania is 1 January 2006. Although a shorter timescale would considerably simplify the economic analysis by reducing uncertainties over policy and economic conditions in both the EU and Romania, accession earlier than the mid-2000s seems unrealistic. On the other hand, a later date would increase the uncertainties to an excessive extent. Over the chosen pre-accession period, the EU's Agenda 2000 (European Commission 1998) and the WTO Uruguay Round Agreement on Agriculture (URAA) present a reasonably reliable framework.

The main basis of the qualitative analysis concerns prospective Romanian agricultural policy, and the CAP itself, as it is likely to exist after implementation of Agenda 2000,[2] and after accession by Romania and other Central European countries. Quantitative estimates of accession effects are calculated using a spreadsheet computer model covering most of the agricultural products of major interest to Romania.

Using projected market analyses for each product, and taking into account policy measures such as production quotas, domestic price supports, and border taxes, the model calculates two alternative market positions for the year of accession, 2006. By comparing these non-accession and accession positions, differences can be calculated in prices and quantities (production, consumption, net trade). Differences in the real incomes (economic welfare levels) of producers and consumers for each commodity represent the costs and benefits of accession to these groups. Further estimates are made of state budget expenditures under non-accession and accession situations in the year 2006, the difference being the effect of accession on the Romanian taxpayer. Together with the producer and consumer effects, a net national economic effect (net social cost or benefit) of Romanian adoption of the CAP can then be calculated.

Model Assumptions and Simulations

Formally, the model represents a partial-equilibrium, comparative-static economic welfare analysis, using conventional neoclassical concepts of price-sensitive supply and demand behavior in a number of agricultural and food product markets. That is, for each commodity, a numerical relationship (equation) relates production or utilization levels to price levels and other factors by means of elasticity or response coefficients. Furthermore, the effects of productivity increases on supply, and of consumer incomes on demand, are estimated using time-trend or shift coefficients.

The model includes fifteen separate farm commodities or first-stage processed food products, viz:

- wheat; maize; sunflower;
- potatoes; sugar; tobacco; wine;
- milk; butter; cheese; skimmed milk powder (SMP);
- beef; pork; poultry; eggs.

No formal connections are assumed between these markets, except for the quantities of milk products (butter, cheese, and SMP), which are derived from total milk production after accounting for liquid milk use. In cases where the farm-gate output immediately undergoes substantial processing (e.g., sugar, tobacco, and milk products), standard physical conversion coefficients are used.

Taken together, these products accounted for over 65 percent of total Romanian farm output in 1997 (see table 11.1), and thus policy analyses for this group should include most of the major effects of interest in EU accession negotiations. However, there are some important exclusions, notably fruits and vegetables, for which EU market legislation exists, although it only involves limited price support. The fisheries and forestry sectors are also excluded.

Table 11.1 Final Agricultural Output and Product Shares, Romania, 1997

Product Group	Value of Output (billion lei)	Product Share %
Wheat	6,592	8.4
Maize	11,620	14.8
Sunflower	982	1.3
Sugar beet	433	0.6
Potatoes	4,712	6.0
Tobacco	139	0.2
Vegetable products included	24,478	31.2
Vegetable sector total	**49,399**	**62.9**
Milk (and milk products)	10,936	13.9
Beef	2,339	3.0
Pork	6,389	8.1
Poultry meat	3,732	4.8
Eggs	3,217	4.1
Animal products included	26,611	33.9
Animal Sector-total	29,113	37.1
Final agricultural output	**78,512**	**100.0**

Note: Wine is classed as an industrial product; its 1997 base output value was lei 14,347 billion.

Source: National Commission for Statistics (2000)

The model results are reported for four simulations, one with a set of standard assumptions, and three "alternatives," each compared with the same non-accession situation. These four simulations are:

Simulation 1: Standard Assumptions

This simulation assumes "normal" values (for developed countries) for rates of growth in farm productivity and GDP (incomes). These values may be overoptimistic in terms of recent Romanian performance, and so would require better agricultural and macro-economic achievement over the next few years until the assumed accession date of 2006. The timetable for EU accession/CAP adoption is now perhaps also optimistic, but avoids speculation about further CAP reform decisions, and should not greatly affect the balance of results. Specifically, the main standard assumptions are:

- farm level productivity for fifteen products in Romania rising on average by around 2 percent per year during the pre-accession period;
- 3 percent per year growth in Romanian GDP and consumer incomes;
- full and immediate adoption of CAP in 2006 according to Agenda 2000 but *without* direct payments to Romanian farm producers;
- CAP quotas for Romanian sugar, tobacco, and milk production set at average pre-accession levels over the period 2001–2005.

Simulation 2: Doubled GDP Growth

This alternative simulation represents a more optimistic forecast for the Romanian economy as a whole, and affects food consumption levels through income-dependent levels of demand rather than farm production or (directly) the agricultural budget although, in practice, a more prosperous macroeconomy would undoubtedly affect farming (e.g., through capital availability) and state finances:

- As Simulation 1, but with 6 percent per year growth in both Romanian GDP and consumer incomes.

Simulation 3: Full EU-15 Direct Payments

This alternative simulation represents a very favorable CAP adoption scenario as regards direct payments. At the time of writing, the European Commission is offering limited direct payments to accession countries during an initial ten-year period of EU membership (*Agra Europe*, May 2002):

- As Simulation 1, but with full (100 percent) application of CAP direct payments at EU-15 levels. These payments, made per hectare or per animal, are currently a major component of EU-15 CAP expenditure, and their extension to farmers in applicant states would considerably increase potential benefits and budgetary costs.

Simulation 4: Lower Quotas

This alternative assumes that the EU is unwilling to accept all the increase in Romanian production that is projected to take place over the pre-accession period, and so imposes production or marketing quotas on Romanian farm production which do not allow farm expansion on accession. This may again be on the optimistic side: at time of writing, the EU has opened negotiations with rather lower quota offers to accession countries:

- As Simulation 1, but with lower CAP quotas, set at 2001 (trend) levels.

Pre-accession Projections

Each of these simulations implies a set of CAP-dependent price levels in 2006 for the modeled agricultural and food products, and for some products a set of quota and direct-payment levels also, as well as consumer income levels. To calculate the effects of these prices compared to their non-accession equivalents on the levels of production and consumption, and hence economic welfare, base-level values for 1997 were first determined. For crops, including sugar, tobacco, and wine, for which yields have varied very greatly due to climatic and structural influences, these values were taken as a five-year average based on production levels since 1993. For livestock commodities, base-level 1997 production is represented by actual recorded "effective production" in that year. For consumption, actual levels of utilization (production minus exports plus imports, ignoring stock changes) for 1997 were used.

Production of each commodity in the year 2006 at constant real prices was then projected by applying a constant commodity-specific annual rate of productivity gain to the 1997 base level. These assumed rates of productivity increase were determined on the basis of expert judgment, taking into account both current Romanian yield levels and experience in more developed agricultural economies. Consumption in Romania has been affected more by macroeconomic instability (see chapter 1) and was projected by applying commodity-specific income elasticities and the standard assumption about GDP growth, or, more precisely, of disposable personal or household incomes.

Following these constant-price projections, the next step was to calculate the change in real prices between the base year 1997 and the accession and non-accession situations in 2006. For the base year 1997, the internal price for some of the main Romanian agricultural commodities (wheat, maize, sunflower, sugar, milk, beef, pork, and poultry) was calculated using the world market price and nominal protection coefficients (NPCs) calculated by Gorton and Deaconescu (1998). These coefficients generally indicated a much smaller gap between external prices and domestic market prices (the latter being lower) than indicated by the formal tariffs. For other commodities (potatoes, tobacco, wine, SMP, cheese, butter, and eggs), NPCs were not available. In some of these cases, a published tariff was used, while in others an NPC value for a related product was used.

For the non-accession scenario in 2006, internal prices were based on the published 1998 Romanian import duties. No changes in these tariff rates were assumed, despite pressures to alter them, upward by agrarian interests in Romania, and downward by external advisors such as the World Bank and the OECD (see OECD 1998 and 2000). Instead, the published tariffs were assumed to become considerably more "operative," so that domestic market protection rises from NPC levels toward the full rate implied by the official tariffs. However, these duties are not liable to be paid on many imports, being supplied under preferential arrangements with certain trading partners. Accordingly, these rates were reduced (usually by 50 percent) to arrive at the actual market protection rate, applied to OECD forecast "world" prices for 2003, using an exchange rate of $1 = Euro 0.9.

For the accession scenario, it was assumed for certain products that the Romanian prices will be the CAP prices proposed under Agenda 2000. For others (e.g., cereals), these support prices are so low that the EU and Romanian market prices will normally be higher, and the world trading price is applicable. For products (e.g., potatoes, pork, poultry, eggs, etc.) where Agenda 2000 provisions do not exist, the EU price was taken as equal to the 2006 estimated world price, or to that price plus an assumed effective EU protection rate. Under these assumptions, Romanian entry into the EU will mean lower rather than higher market protection for many farm products.

Using these price changes, supply and demand price elasticities were applied to convert the constant-price projections to final 2006 projections (tables 11.2 and 11.3). The assumed price elasticities represent fairly modest adjustments, and are not far from values commonly used in modeling agriculture in the EU, the United States, and other developed countries. However, given the substantial changes in prices assumed over the pre-accession period, they imply considerable changes in production and consumption levels.

Welfare Results

These projection results for prices and quantities determine producer and consumer welfare effects according to standard economic concepts of changes in producer and consumer "surplus." Finally, taxpayer welfare effects are derived from the implications for both national expenditures on agriculture, and for flows to and from the EU budget (see below), again comparing accession with non-accession in the year 2006. The calculations for the standard scenario were repeated, with appropriate modifications, for each of the above-specified alternatives.

Table 11.4 shows the economic welfare (real income) effects on producers and consumers, product by product. For the modeled crop sectors, the main effects are seen in the cereal sectors of wheat and maize, with similar though much smaller effects for sunflower. For these sectors, with the standard set of assumptions, producers suffer income losses, since accession enforces the elimination of Romanian market protection for grains. Under the proposed CAP market regimes for these products, arable farmers in Romania will lose all protection from competition from efficient producers in the EU-15 (and other new member states such as Hungary). On the other hand, users of cereals, who include farmers feeding grain to livestock, would benefit from the lowering of prices at accession. Given that Romania is a surplus producer for these products, the net welfare effect of accession for these arable sectors is negative (ignoring budgetary aspects).

Table 11.2 Base and Projected Production Levels, Romania, 1997 and 2006

	Units	Base Level, 1997	Assumed Annual Productivity Growth, 1997–2006	Assumed Supply Elasticit y	Projected Level, 2006*	Percentage Change from Base
Wheat	thou. t	6,133	2.0%	0.8	8 521	39%
Maize	thou. t	9,690	2.0%	0.8	11,400	18%
Sunflower	thou. t	945	2.5%	0.8	1,200	27%
Potatoes	thou. t	3,343	2.0%	1.2	4,217	26%
Sugar	thou. t	275	3.0%	1.2	404	47%
Tobacco	thou. t	21	3.0%	1.2	30	43%
Wine	thou. hl	6,612	1.0%	1.2	8,107	23%
Milk	thou. hl	53,595	1.5%	1.2	62,734	17%
SMP	thou. t	11	1.5%	1.2	14	28%
Cheese	thou. t	206	0.5%	1.2	263	28%
Butter	thou. t	9	1.5%	1.2	11	34%
Beef	thou. t	220	1.0%	1.0	169	–23%
Pork	thou. t	540	1.0%	1.0	524	–3%
Poultry meat	thou. t	153	1.0%	1.5	111	–28%
Eggs	millions	4,983	1.0%	1.5	4,671	–6%

Note: thou. = thousand; t = metric tonnes; hl = hectoliters. Livestock units are deadweight.

* Projected, taking into account (a) productivity growth and (b) change in real domestic price levels (not shown) between 1997 and 2006 non-accession

Table 11.3 Base and Projected Consumption Levels, Romania, 1997 and 2006

	Units	Base Level, 1997	Assumed Income Elasticity	Assumed Demand Elasticity	Projected Level, 2006*	Percentage Change from Base
Wheat	thou. t	5,760	0.2	–0.1	6,074	5%
Maize	thou. t	8,784	0.2	–0.1	9,262	5%
Sunflower	thou. t	920	1.0	–0.3	1,179	28%
Potatoes	thou. t	3,294	0.0	0.0	3,294	0%
Sugar	thou. t	403	1.0	–0.3	525	30%
Tobacco	thou. t	25	1.5	–0.8	37	48%
Wine	thou. hl	5,635	1.2	–1.2	7,696	37%
Milk	thou. hl	22,084	1.5	–0.2	28,565	29%
SMP	thou. t	10	1.5	–0.5	14	40%
Cheese	thou. t	207	1.0	–0.5	270	30%
Butter	thou. t	9	1.5	–0.5	12	33%
Beef	thou. t	213	1.5	–0.7	310	46%
Pork	thou. t	500	1.0	–0.5	652	30%
Poultry meat	thou. t	269	1.0	–0.5	351	30%
Eggs	millions	4,370	1.0	–0.5	5,702	30%

Note: thou. = thousand; t = metric tonnes; hl = hectoliters. Livestock units are deadweight.

* Projected, taking into account (a) annual GDP growth at 3 percent and (b) change in real domestic price levels (not shown) between 1997 and 2006 non-accession

Table 11.4 Producer and Consumer Welfare Effects by Products and by Model Simulations (billion lei, 1997 prices)

	Standard Assumptions		Doubled GDP Growth		Full EU-15 Direct Payments		Lower Quotas	
	Prod	**Cons**	**Prod**	**Cons**	**Prod**	**Cons**	**Prod**	**Cons**
Wheat	–1,059	780	–1,059	820	2,133	780	–1,059	780
Maize	–1,558	1,325	–1,558	1,393	3,485	1,325	–1,558	1,325
Sunflower	–160	166	–160	203	331	166	–160	166
Potatoes	0	0	0	0	0	0	0	0
Sugar	613	–983	613	–1273	613	–983	559	–983
Tobacco	62	–82	62	–114	62	–82	53	–82
Wine	925	–742	925	–993	925	–742	925	–742
Crops	*–1,177*	*464*	*–1,177*	*36*	*7,549*	*464*	*–1,239*	*464*
Milk	965	–343	965	–343	2,321	–343	924	–327
Butter	n.a.	2	n.a.	–48	n.a.	–35	n.a.	–35
Cheese	n.a.	–670	n.a.	–868	n.a.	–670	n.a.	–670
SMP	n.a.	–35	n.a.	3	n.a.	2	n.a.	2
Beef	–38	91	–38	127	2,012	91	–38	91
Pork	–1,339	2,143	–1,339	2,775	–1,339	2,143	–1,339	2,143
Poultry meat	–250	1,165	–250	1,508	–250	1,165	–250	1,165
Eggs	–389	611	–389	791	–389	611	–389	611
Livestock	*–1,051*	*2,963*	*–1,051*	*3,944*	*2,355*	*2,963*	*–1,092*	*2,980*
Total	*–2,228*	*3,427*	*–2,228*	*3,980*	*9,904*	*3,427*	*–2,331*	*3,444*

Note: Prod. = producers; Cons. = consumers; n.a. = not applicable

The effects of accession for potatoes and tobacco are small or zero, although for tobacco the EU regime of production quotas and high prices benefits Romanian producers and penalizes consumers. For sugar, a similar regime leads to the same but much larger economic effects for producers and consumers. For wine, even assuming some rise in Romanian market protection before accession, the CAP brings income gains to producers and losses to consumers. For the seven crop sectors as a whole, the net effects of accession under the standard assumptions involve an income cost of nearly lei 1,200 billion (at 1997 prices) for Romanian producers, somewhat offset by a gain of lei 464 billion for the consumers of these products. Accession threatens the economic welfare of extensive-crop arable producers and of consumers of sugar, wine, and tobacco, but these losses are partly offset by real-income gains to consumers and users of cereals, and to quota holders of the three "intensive" crops.

For the livestock sectors (dairy and beef cattle, pigs, and poultry, covering eight livestock products), again a mixture of results is apparent. Dairy producers, even without direct payments proposed in Agenda 2000, stand to benefit from adopting the CAP quota/price regime, while consumers of fresh and processed milk products, especially cheese, suffer real-income losses from the higher prices involved. Given the considerable lowering of domestic market protection for Romanian beef assumed for the pre-accession period, only small economic effects of accession are seen for this sector. For the pig and poultry sectors, even assuming some lowering of domestic market protection before accession, considerable losses to producers (offset somewhat by cheaper feed grains mentioned above) are more than offset by real income benefits to Romanian consumers of pork, poultry, and eggs. For the modeled livestock sectors as a whole, Romanian EU Accession in 2006 (under standard assumptions) involves a loss of about lei 1,050 billion in real producer income, slightly smaller than for the crops sector, but substantial real-income gains of nearly lei 3,000 billion to consumers of livestock products.

The overall annual economic effect on Romanian farm producers of accession under standard assumptions is estimated as a real-income loss of lei 2,228 billion, an average loss of lei 743,000 per farmer. As explained above, there are positive as well as negative economic impacts among the various sectors. For Romanian consumers (including farm users of feed crops), the net real-income gain is about 50 percent larger, at lei 3,427 billion, representing about lei 150,000 per citizen annually.[3]

Higher economic growth in the Romanian economy during the period before accession (Simulation 2) affects only the consumer side of the calculations. In general, the change is positive, since the higher incomes would be (partly) spent on food. Thus, the cost savings from generally lower market protection under the CAP at accession compared to non-accession would increase the economic gains from this change in policy. The estimated increase in consumer gains is about lei 550 billion, an amount which increases the net effect of accession on national income (GDP) by about 50 percent.

Payment of full EU-15 direct payments to Romanian farm producers as a component of Accession (Simulation 3) would completely alter the outcome for many farmers, converting the standard-assumption loss of lei 2,228 billion into a real-income gain of lei 9,904 billion. EU taxpayers must bear the costs of this expenditure, but Romania's share of the increased spending (in the ten CEECs) is relatively small, leading to a net Romanian taxpayer cost of Accession of just over lei 200 billion.

Finally, application of more severe quota limitations on Romanian producers of sugar, tobacco, and milk (Simulation 4) has relatively small additional economic effects. The output foregone in this scenario has presumably the highest (marginal) production costs, and therefore only small amounts of profits (producer incomes) are withheld. There are small effects on consumers arising from the necessary reductions in levels of output and perhaps consumption of milk and milk products, but the net effects are negligible.

In order to calculate the net effect on national economic welfare (national income, or GDP), it is necessary to consider the effect of accession on public expenditure, that is, on taxpayers and the state budget. Accession will mean that Romania must contribute to the overall EU budget of which about half (Euro 40 billion, or lei 315,400 billion) is CAP expenditure. Using Romania's share of the GDP of an enlarged EU, that is, about 0.4 percent, the budget or taxpayer cost of accession is lei 1,257 billion. This, of course, is a loss to national income.

Further, the CAP involves some so-called structural expenditure from the European Agricultural Guidance and Guarantee Fund (EAGGF) Guidance section of the EU budget. Romania's share of this cost may be considered as included in the Euro 40 billion figure above. More important, such expenditure is normally dependent on cofinancing from national budgets. The level of EAGGF Guidance expenditure in Romania in the year 2006 is almost impossible to estimate directly, but an approximation may be Romania's share of pre-accession SAPARD funding, which is some Euro 150 million per year (National Agricultural Rural Development Plan 2000–2006). Using this figure (equivalent to lei 1,183 billion at the 1997 exchange rate), and a cofunding rate of 25 percent, an additional cost to the Romanian national budget can be estimated at lei 300 billion.

Finally, accession will mean that certain national agricultural subsidies must be phased out, while others will have to be modified in line with EU rules on state aids. A rough estimate is to assume a reduction (budget saving) of half the 1997 expenditure on the voucher scheme. This produces a budget gain of some lei 1,400 billion.

Together, these three budgetary or taxpayer effects of accession largely offset each other, leaving a negative net taxpayer effect of accession under standard assumptions of about lei 153 billion. Thus, Romanian Accession (for agriculture alone) under these conditions appears almost (state) budget-neutral.

The three components of change in the level of national economic welfare are producer effects (net losses in aggregate), consumer effects (net gains), and taxpayer effects (a small gain). Taken together, these components result under standard assumptions in a net gain in Romanian national income of some lei 1,042 billion, or lei 45,000 per citizen (table 11.5 and figure 11.1). Given the GDP figures reported above, this represents a once-for-all rise of about 5 percent in 1997 Romanian GDP, or, given economic growth in the intervening period, about 4 percent in the year 2006.

As regards the alternative Simulations 1 to 3:

- Doubled GDP/income growth during the pre-accession period increases consumer gains in 2006 since more food is then being consumed, but does not significantly affect production or taxpayer cost. Compared to the Standard Assumptions result, the positive net effect rises by about 50 percent to lei 1,595 billion, or lei 69,000 per citizen.

Table 11.5 Economic Welfare Effects of EU Accession, Alternative Scenarios (billion lei)

	Standard Assumptions	Doubled GDP Growth	Full EU-15 Direct Payments	Lower Quotas
Producers	–2,228	–2,228	9,904	–2,331
Consumers	3,427	3,980	3,427	3,444
Taxpayers	–157	–157	–345	–157
Net	1,042	1,595	12,986	956
(per citizen—th. lei)	(45)	(69)	(564)	(41)
Differences				
Producers	–	0	12,132	–103
Consumers	–	553	0	17
Taxpayers	–	0	–188	0
Net	–	553	11,944	–86

Note: th. = thousand

- Full direct payments have a huge impact: Romanian farm producers gain almost lei 10,000 billion, while taxpayers have to contribute about lei 345 billion to the state and EU budgets, instead of lei 157 billion. The net effect on national economic welfare is lei 12,986 billion, a huge gain of nearly lei 12,000 billion over the standard-assumption effect.
- Lower production quotas for milk, sugar, and tobacco limit accession gains for these producers, and hence slightly increase the overall producer loss. The net effect on national income is a little less than from the Standard Assumption simulation.

Figure 11.1 Economic Welfare Effects of EU CAP Adoption, Romania: Alternative Scenarios

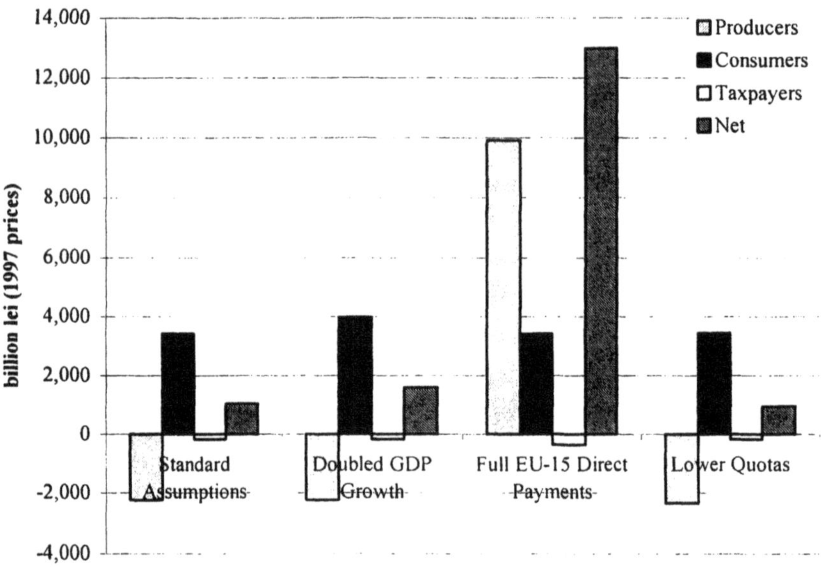

In all four simulations of the model, the net change in national economic welfare due to accession is a gain, bigger or smaller, according to the different assumptions. Romanian consumers gain between lei 3,400 and 4,000 billion, depending on general economic growth, while farm producers will lose about lei 2,300 billion unless direct payments are forthcoming. Romanian taxpayers may gain or lose a relatively small amount (and from higher income levels), depending on adjustments to national subsidy systems and on the overall cost of the EU-25 CAP.

Conclusion

The aim of this study was to estimate the major economic costs and benefits related to Romanian accession to the CAP of the EU, calculated as the differences between the "non-accession" and "accession" situations in the assumed accession year of 2006 (not changes since the base year of 1997). The model results are more strongly influenced by the assumptions about the policy conditions in the year of accession than by those concerning the intermediate period.

There are several main findings from the study. Without direct payments, the producers of wheat, maize, and sunflower bear significant economic welfare losses; other crop products are less sensitive to the specific assumptions of the model. For vegetable products other than potatoes, and for wine, producers gain and consumers lose. For crop products subject after accession to a quota regime, that is, sugar and tobacco, the effects of different assumptions as to the levels of these quotas on producer welfare are relatively small. For the livestock sector, producers bear a welfare cost of approximate lei 1,000 billion, while the net effect on consumer welfare is a gain of lei 3,000–4,000 billion, which is, however, sensitive to the level of real consumer incomes. Milk producers gain in all simulations, especially in the case of full direct payments. For meat and eggs, consumer gains exceed producer losses, except for beef with direct payments, which maximize gains, shared by both producers and consumers. Apart from direct payments, the main economic commodity issues for Romania in the accession negotiations concern cereals, meat, and milk.

Notes

1. The model that this chapter reports was constructed by Agricultural Reform and EU Alignment (AREA) Group staff during the period 11 May to 11 November 1998 as part of PHARE project no. RO-9505-04-01. Responsible Group members included Carmen Firici, Jacqueline Leonte, Silviu Balanica, Mihaela Preda, and Victoria Burtea. The full report was submitted to the Ministry of Agriculture and Food (MoAF).

2. At the time when the research for this chapter was carried out, only the July 1997 and March 1998 proposals of the European Commission

were available. Although these proposals were largely approved by the Council of Agricultural Ministers in February 1999, there were some changes introduced in the council's decision.

3. This may be compared with 1997 GDP of lei 1,083,000 per head.

References

Agra Europe. No. 2005. 31 May 2002

European Commission. 1998. *Agenda 2000—Commission Proposals*. Brussels, Belgium: European Commission.

Gorton, M. and Deaconescu, D. 1998. *Domestic Resource Costs and International Competitiveness of Romanian Agriculture*. PHARE Project EU Alignment and Policy Reform Report. Mimeo-graphed.

National Commission for Statistics. 2000. *Anuarul statistic al României*. Bucharest, Romania: National Commission of Statistics.

Organization for Economic Cooperation and Development. 1998. *The Agricultural Outlook 1998-2003*. Paris: Organization for Economic Cooperation and Development.

—— 2000. *Review of the Agricultural Policies: Romania*. Paris: Organization for Economic Cooperation and Development.

Chapter 12

WTO Commitments and CAP Adoption in Romania: Model Results

Sophia Davidova, Daniela Giurca, Mirela Rusali, and Lionel Hubbard

Three simultaneous processes shape agri-food policies in the CEECs: their domestic reforms; preparation for accession to the EU; and external trade commitments, both to the WTO and under various regional and bilateral agreements. These processes are interrelated, which raises difficult issues for policymakers who have to foresee the multidimensional effects of their decisions. The situation is likely to become more complicated as preparations for accession to the EU intensify and as the WTO negotiations on agriculture begin.

Various studies have examined the issue of interdependence between the adoption of the CAP and the CEECs' WTO commitments in agriculture (Tangermann and Josling 1994; Buckwell et al. 1995; Mahé 1996), as well as the general effects of the Uruguay Round Agreement on Agriculture (URAA) on the CEECs and the Former Soviet Union (OECD 1994a, b, and c). However, most of these studies either treat the CEECs as a single region, or focus on the Central European countries, which are most advanced in their preparation for accession. Relatively little research has been undertaken on Romania, which is not a frontrunner for accession.

Romania is an interesting case, first, because of the size and importance of agriculture in the national economy, and second, because of its unique situation among the EU applicant countries of having developing country status within the WTO. This chapter addresses two main questions:

- Will Romanian agricultural policy be constrained by the (assumed) new commitments resulting from the new WTO negotiations on agriculture?
- Will the Romanian WTO commitments raise problems for the EU, which might result in sensitive areas during the negotiations for accession?

The results reported are on the likely situation in Romania in 2006, which is taken to be a notional year for accession. Projections under accession and non-accession scenarios are generated using the model reported in chapter 11.

URAA and Romanian Commitments

Contracting Party Commitments under the URAA

For the first time, the Uruguay Round (UR) of GATT agreed stringent disciplines on agricultural trade in four areas: market access, domestic support, export subsidies, and sanitary and phytosanitary measures. The fourth area is not analyzed here. Because of its crucial importance for the development of fair trade in agri-food products, the UR has been defined as a "milestone for the multilateral trading system" with a role "to guide the trading world into the new millennium" (Josling, Tangermann, and Warley 1996, 133). The URAA provides for different commitments for developed countries and for developing countries. These are summarized in table 12.1. Romania is a unique case among those CEECs which are WTO members, in that it has the status of a developing country.

In the field of market access, all GATT contracting parties agreed to convert all non-tariff barriers (e.g., quantitative import restrictions, variable import levies, minimum import prices) into tariffs equal to the difference between the average 1986–1988 domestic price and the world market price for the same period. This "tariffication" aims at greater transparency and predictability of world agricultural trade. It is coupled with the commitment to "bind" (put a ceiling on) all tariffs. Thus, maximum tariffs, so-called tariff bindings, are known for each agricultural product in each participating country. Under the

Table 12.1 Structure of the Uruguay Round Agreement on Agriculture

Market Access (base period 1986–1988)		Export Subsidies (base period 1986–1990)		Domestic Support (base period (1986–1988)	
Developed Countries	**Developing Countries**	**Developed Countries**	**Developing Countries**	**Developed Countries**	**Developing Countries**
Tariffication of non-tariff barriers. Reduction of the tariff bindings by 36 percent on average over six years. Minimum reduction of individual tariff lines by 15 percent. Minimum access commitments for products subject to tariffication: 3 percent of domestic consumption in 1986–1988, rising to 5 percent over six years. Maintenance of current access opportunities	Tariffication of non-tariff barriers. Reduction of the tariff bindings by 24 percent on average over ten years. Minimum reduction of individual tariff lines by 10 percent. Minimum access commitments for products subject to tariffication: 3 percent of domestic consumption in 1986–1988, rising to 5 percent over ten years. Maintenance of current access opportunities	Product specific reductions of budgetary outlays for export subsidies in comparison to the base period by 36 percent over six years. Quantities exported with subsidies to be decreased by 21 percent over six years from 1986–1990 base.	Product specific reductions of budgetary outlays for export subsidies in comparison to the base period by 24 percent over ten years. Quantities exported with subsidies to be decreased by 14 percent over ten years from 1986–1990 base.	Domestic support that is non- or minimally trade distorting is put in the "green box" and is exempt from reduction. Direct payments under production limiting programs are put in the "blue box" and are exempt from reduction. Distortive domestic support is measured by total Aggregate Measure of Support (AMS) across all commodities ("amber box"). Base total AMS referring to 1986–1988 to be decreased by 20 percent over six years in equal annual installments. Distortive support is exempt from reduction if it does not exceed 5 percent of the value of a product or of the total value for agricultural production (de minimis).	Domestic support that is non- or minimally trade distorting is put in the "green box" and is exempt from reduction. Direct payments under production limiting programs are put in the "blue box" and are exempt from reduction. Distortive domestic support is measured by total Aggregate Measure of Support (AMS) across all commodities ("amber box"). Base total AMS referring to 1986–1988 to be decreased by 13 percent over ten years in equal annual installments. Distortive support is exempt from reduction if it does not exceed 10 percent of the value of a product or of the total value for agricultural production (de minimis).

Source: author's compilation following OECD (1995)

URAA, developed countries had to decrease these tariff bindings by an average of 36 percent over a six-year period (1995–2000). The commitment for developing countries is to decrease the bindings on average by 24 percent over ten years (1995–2004). The cut in individual tariff lines can be higher or lower than the average, but the minimum cut is 15 percent for developed countries and 10 percent for developing countries. The only possible direction of future WTO negotiations is to further decrease these bindings.

Countries also have to open low tariff quota, under so-called current access and minimum access. The former means that opportunities for imports should be at least as favorable as those prevailing in the base period (1986–1988). The latter is to ensure that there are opportunities for imports equal to at least 3 percent of domestic consumption, rising to 5 percent in the final year of implementation. However, countries can make use of a special safeguard provision. According to stringent rules, this allows the imposition of a temporary additional import tax in cases of a sudden increase in imported quantities or decrease in import price.

As the objective of URAA was to start the process of removing distortions to world agricultural trade, contracting parties have also accepted commitments in their domestic policies. Trade-distortive policy measures are in the so-called amber box and are subject to measurement and reduction. The indicator for their measurement is the AMS. Countries, including Romania, which did not calculate AMS explicitly are allowed to apply "de minimis" distortive support. This is also expressed in value terms. For the developed countries it is equal to 5 percent of the value of production of each product for product-specific measures, or to the same percentage of the total value of agricultural output for non-commodity-specific measures. For developing countries, the percentage is 10 percent.

The non- or minimally distortive measures, which are included in the "green box," can be applied by the national governments without any external constraints. These are support measures that do not affect trade or production. For developed countries they include: research, training, and extension services; pest and disease control; marketing and infrastructure services; public storage for food security reasons; decoupled income support; calamity payments; payments under environmental programs; and regional aid for less-favored areas. For developing countries the list also includes investment and input subsidies.

The "blue box" of measures is the result of a compromise between the EU and the United States and relates to "direct payments under production-limiting programs." Thus, acreage and headage payments in the EU and deficiency payments in the United States are exempt from commitments to decrease support. However, not all the payments included are related to production-limiting programs. According to the URAA, measures not decoupled from production, and thus not neutral to trade, are allowed if they do not pertain to a production level above 85 percent of the output in the base period. For this reason it is expected that the forthcoming negotiations on agriculture will focus on measures in the blue box. Several WTO negotiating countries with liberal agricultural policies, such as New Zealand, Canada, and Australia, would like the removal of the blue box and allocation of policy measures actually in this box to either the amber or green box. This allocation will depend on whether the support measures are minimally trade distortive or not. It is difficult for the EU to claim that its direct payments, including the ones defined in the Agenda 2000 decisions, are really decoupled from production, that is, that they are nondistortive. The EU trade commissioner has underlined that the blue box "would be a prime target for the EU's trading partners in the next trade round, and a key priority for the EU to defend" (*Agra Europe*, January 1999).

During the UR, it was particularly difficult to achieve a compromise over cuts in the quantities of agricultural products that receive export subsidies and in the value (budgetary outlay) spent on these subsidies. This is because export subsidy commitments induce the strongest constraint on domestic agricultural policies. However, an agreement was finally achieved. Budgetary outlays have to be reduced by 36 percent for developed countries, and by 24 percent for developing countries, from the amounts during the base period, 1986–1989. Again, reductions were to be made by equal annual installments over six years (1995–2000) for developed countries and ten years (1995–2004) for developing countries. Quantities exported with subsidies were to be reduced by 21 percent for developed countries, and 14 percent for developing countries, over the same periods (see table 12.1). Reductions were made on a commodity-specific basis. Developing countries were deemed exempt from reduction commitments on subsidies that reduce marketing costs of agricultural exports and from subsidies on internal transport and freight shipments.

The Agricultural Commitments of Romania

Romania set its own commitments on the basis of the above rules. Its developing country status enabled Romania to set high tariff bindings for agricultural products. With the exception of agricultural input items, namely, seeds and purebred animals, Romania used fifty-seven different combinations of levels of initial binding and tariff cuts in its agricultural schedule. The maximum initial binding was set for pork (370 percent), and the minimum for meat of sheep and goats, live poultry, and some live plants (20 percent). The weighted average was 161.5 percent (Tesliuc 1996). The tariff cuts varied between 10 percent, the minimum necessary under the URAA, and 88 percent for ivory, a product of no trade importance for Romania. Adopting a strategy similar to that of developed countries, Romania made larger cuts in the lower tariffs (for discussion about this practice, see Tangermann 1996). For example, the initial binding for pork at 370 percent was cut by Romania by the minimum of 10 percent only. Thus, the resulting final binding was 333 percent. On the other hand, the initial binding for legumes of 75 percent was cut by 38 percent, giving a final commitment was 49 percent.

Romania does not need such high bindings in order to protect its agriculture: the bindings contain both "traditional water" and "policy water." That is, there are large differences between nominal protection, the published tariffs (that is, the tariffs actually applied), and the tariff bindings. Protection expressed in the form of nominal protection coefficients (NPCs),[1] is rather low and is consistently lower than the tariffs published in 1997, while these are far lower than the final year (2004) tariff bindings. Table 12.2 illustrates the difference between published tariffs and the final bindings for some main agricultural products. The tariff bindings place no real constraints on Romanian agricultural policy.

Table 12.2 Tariff Bindings and Published MFN Tariffs in Romania, 1997

Commodity	Final Bindings (%)	Published Tariffs (%)
Wheat	276	25
Maize	276	30
Sunflower	184	30
Sugar	192	25
Milk	141	35
Pork	355	60
Beef	336	50
Poultry	134	60

Source: Romanian Customs Tariff (1998)

Romania has opened eleven minimum access quotas (table 12.3). With the exception of seed potatoes, all are for processed foods, cigarettes, and alcohol. The in-quota quantities are relatively small, while the in-quota tariff rate is high, around 100 percent. For this reason, minimum access is not explicitly included in this analysis, as it is assumed that it cannot greatly affect the tradable surplus.

Table 12.3 Current and Minimum Access Quotas ('000 t)

Products	Current Access	Consumption	Minimum Access: Initial	Minimum Access: Final
SMP	none	17.3	0.5	0.9
Cheese	none	52.0	1.6	2.6
Seed potatoes	11.7	300.0	0	3.3
Chocolate	0.5	23.9	0.2	0.7
Pasta	none	78.9	2.4	3.9
Ice cream	none	15.1	0.4	0.7
Beer*	20.0	14,008.0	400.0	680.0
Vermouth*	0.3	15.0	0	0.5
Undenatured ethyl alcohol: 80% vol. or higher	none	n.a.	0	0.5
Undenatured ethyl alcohol: less than 80% vol.	3.8	269	4.3	9.6
Cigarettes	0.2	32.6	0.8	1.4

* thousand hl.
n.a. = not applicable
Source: Uruguay Round LXIX Schedule-Romania

As was the case with most developing countries, Romania did not record the base level of domestic support (AMS) in its GATT schedule, but used the de minimis clause. This practice has been described as a weakness which puts developing countries at a disadvantage, as they still need output-enhancing policies (Josling 1996). It is even more serious in the case of Romania, faced with the prospect of EU membership and adoption of the CAP.

The list of allowed export subsidies is short, but it incorporates the most important agricultural products, including some of the products for which the EU uses export subsidies (table 12.4). Romanian export subsidies (budgetary outlays) are expressed in constant 1986–1989 lei, and are thus protected from erosion by inflation. However, the budgetary allowances are far lower than in the other CEECs that are major agricultural producers. According to estimates from the U.S. Department of Agriculture, the final year (2004) commitments for Romania, expressed in 1992 U.S. dollars, account for 147 million, while for Hungary and Poland for year 2000 they are 419 and 500 million, respectively (Plunkett and Maxwell 1994).

Overall, the Romanian commitments seem quite inconsistent. They permit very high tariff protection, but impose tight constraints on (coupled) domestic support, via the AMS commitment, and allow only modest use of export subsidies. This means that Romania cannot make extensive use of the high bindings and is bound to follow relatively liberal agricultural policies. From an economic point of view, this is advantageous, particularly as the country still faces massive restructuring of the agri-food sector, with efficiency and international competitiveness as implied objectives. However, from the point of view of accession to the EU and the adoption of CAP, these constraints on domestic policies might foreshadow difficult issues during the accession negotiations.

Table 12.4 Romanian Export Subsidy Commitments

Commodity	Final Binding, 2004		
	Outlay (mln lei)[b]	Outlay (mln ECU)[c]	Volume ('000 t)
Cereals	494.4	22.9	283.5
Wheat and wheat flour[a]		16.6	209
Maize[a]		3.4	43.5
Sugar	421.4	19.6	147.6
Oilseeds	1.0	0.05	2.75
Sunflower seeds[a]		0.014	0.9
Beef, pork, sheep, and processed meat	389.3	18.1	137.6
Pork[a]		15.4	119.7
Beef[a]		0.9	7
Poultry	167.2	7.7	27.8
Vegetable oil	193.2	8.9	83.6
Sunflower oil[a]		7.4	71.1
Live animals	90.4	4.2	16.7
Cheese	22.6	1.03	11.1
Butter	131.0	6.1	14.5
Wine	50.40	2.3	7.9
Eggs	0.21	0.01	1.2
Fruit	109.20	5.1	132.8
Vegetables	133.50	6.2	111.3

[a] Initial schedule has been disaggregated using the share of the product group in the total export under the relevant chapter in the Harmonized Customs System.
[b] As in the schedule, constant lei 1986–1988.
[c] Recalculated at 1997 lei prices and converted into ECU at 7,885 lei/ECU.
mln = million.
Source: Uruguay Round LXIX Schedule – Romania.

Approach, Assumptions, and Information Sources

Similarly to the time frame in chapter 11, the analysis is focused on the year 2006, under accession and non-accession situations. Three main groups of assumptions are necessary:

- developments in Romania to the year 2006;

- world and EU agricultural prices in the year 2006;
- the outcome for agriculture of the WTO negotiations.

Developments in Romania

Several of the assumptions used here are identical with those of the model of Economic Costs and Benefits of Romanian Accession to the EU (chapter 11) since some of the results from that model are used in the analysis of WTO commitments. Accordingly, the base year data are for 1997, with assumed annual growth rates in Romania through to 2006 of 3 percent for GDP and 2 percent for agricultural productivity. Import flows are maintained at their 1997 levels. This is due to the fact that the erratic trade policy in Romania before 1997 makes it difficult to define a trend in agricultural imports (by mid-1997, Romania had, to some degree, liberalized and stabilized its foreign trade policy). Consequently, the assumption employed is that the growth rate of imports through to 2006 is zero. Romania's real exchange rate is taken to be the average 1997 level.

After the fulfillment of its current UR commitments in 2004, it is assumed that Romania will change its WTO status from developing to developed country, in preparation for EU accession. Policy assumptions under the accession scenario are the same as in chapter 11 and thus include the adoption of EU rates of prices and protection; the implementation of production quotas for sugar, milk, and tobacco; and direct payments. As a reference point for products subject to quota, the situation with unconstrained production is also estimated.

One of the crucial determinants in the analysis is the rate of protection in Romania in the base year and in 2006 under non-accession. Most of the assumptions are the same as in chapter 11. For 1997, for cereals, sunflower, sugar, milk, and meats, NPCs calculated by Gorton and Deaconescu (1998) are used. For dairy products, the preferential duty under the EA is used, and for the remaining products the published tariffs are adjusted to take into account the preferential imports under various trade agreements. However, the level of internal prices is lower than suggested by the published tariffs—due not only to preferential imports, as assumed in chapter 11, but also to the underdevelopment of markets and the low level of market efficiency (Davidova and Ivanova 1997). It is assumed that market efficiency will increase through to 2006, but that price transmission will continue to be imperfect. For this reason, under the non-accession scenario, two alternatives are used in the analysis of AMS and export subsidies: protection rates equal to 0.5 and 0.25 of the 1998 published tariffs. There are two exceptions, milk and tobacco, which are not included in any preferential trade agreement. For these, under the first alternative, the 2006 protection rate is set equal to 1998 published tariffs, and under the second, to half of these tariffs.

The analysis of market access and alignment to the EU focuses on the tariff bindings and not on the published tariffs. The reason for this is that, upon enlargement, third countries are likely to pay attention to the bindings, because they are negotiable within the WTO (see Buckwell et al. 1995). The basic criterion for choosing tariff positions for comparison between Romania and the EU was the share of import value of the respective group in total Romanian agricultural imports. For an indication of the order of magnitude of possible compensation claims, it is assumed that by 2006 the other CEFTA countries will also be EU members. Consequently, they are excluded from the group of third countries.

As the National Commission for Statistics in Romania does not publish aggregate figures for gross agricultural output, in the analysis of AMS, all distortive support was treated as product specific and the de minimis clause was applied product by product.

In the Romanian UR schedule of export subsidy commitments, cereals, oilseeds, and meats (with the exception of poultry) are aggregates. In the present study, these groups are disaggregated according to the share of individual products in the total exports of the respective chapter in the Harmonized Customs System (see table 12.4). Finally, Romania protected its budgetary outlays on subsidized imports against inflation erosion by expressing the commitments in constant lei. The commitments for 2006 were recalculated in 1997 prices using the consumer price index provided by the Romanian National Bank and the average 1997 exchange rate of lei to ECU.

World and EU Agricultural Prices

Projections for world prices are based mainly on OECD or World Bank data. For the EU, the Agenda 2000 prices are used where possible, and for products for which there are no proposals, an EU protection rate in 2006 is assumed. For several products, for example, pork and poultry, the assumed protection is zero (see chapter 11).

WTO Negotiations on Agriculture

The possible outcome of the new WTO negotiation on agriculture has been widely discussed among academics and politicians (e.g., Josling, Tangermann, and Warley 1996; de Zeeuw 1997). The possible options for new commitments in the fields of market access include: a major across-the-board tariff binding reduction, for example, by 50 percent; a formula to reduce the higher tariffs at a greater rate; reduction of the bindings to the average applied tariffs over a multilaterally agreed base period; a "zero for zero" approach for full elimination of tariffs on certain products. Several ideas have been developed about the possible outcome in the field of export subsidies, from total elimination (de Zeeuw, 1997) to continuation of the gradual phasing out of overt export subsidies and imposition of international discipline on covert subsidies through government export credits and export guarantees. Particularly sensitive is expected to be the domestic support issue with ideas for refining the green box, reconsideration of the blue box, and removal of partially decoupled payments from reduction exemption. As for the completion of negotiations and the start of implementation, the opinion expressed by the EU trade commissioner is that the aim is to conclude the negotiations by the end of 2003 (*Agra Europe*, January 1999).

Assumptions used for the current analysis are more conservative and simplified, namely, that:

- the implementation of the new WTO commitments will start in 2004;
- the outcome will be at least equal to what was achieved during the UR;
- the commitments for cuts in the tariff bindings, domestic support, and export subsidies will be the same as in the UR;

- the negotiating round will be more radical in defining distortive domestic support, namely, by including the EU's direct payments in the amber box;
- if upon accession to the EU, direct payments apply to Romania, they will be calculated in the amount of distortive support.

Results

Tariff Bindings and Compensation

Figure 12.1 shows that, in general, Romanian tariff bindings are higher than those in the EU. Thus, compensation might not be a sensitive issue during the negotiations for accession. However, exceptions are sugar, citrus fruit, bananas, and rice, and these are some of the most important items in Romania's agri-food imports (23 percent in 1997).

Figure 12.1 Tariff bindings 2006, Romania and the EU

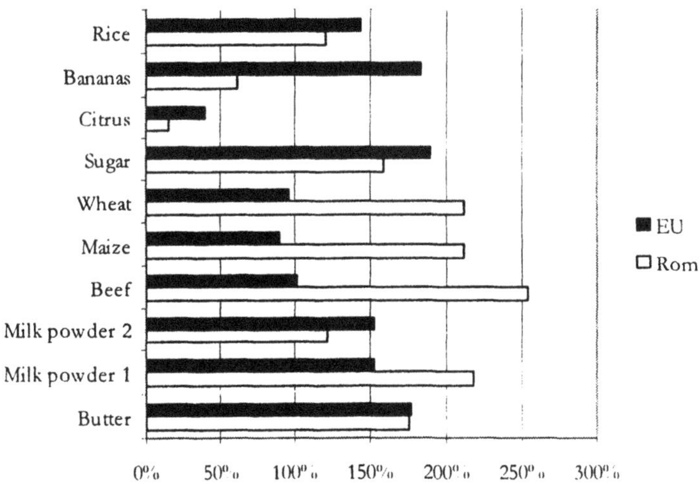

The value of potential compensation to the third country importers is shown in table 12.5. Most interesting is the case of sugar, because approximately one-fifth of the value of imports comes from Moldova and, under the free trade agreement with that country, enters with a zero tariff. In this case, the value of compensation is likely to be higher, because of the large discrepancy with the EU projected binding for 2006. The overall value of compensation for the four products is estimated at $50 million, with the main beneficiary being Moldova ($30 million). However, the total is not substantial, and thus this issue is not expected to create serious problems during the accession negotiations.

Table 12.5 Potential Compensation to Third Countries

NS Code	Product	Romanian Tariff Bindings 2006 (%)	EU Tariff Bindings 2006 (%)	Imports from Third Countries ('000 U.S. dollars)	Compensation ('000 U.S. dollars)
		1	2	3	4*
1701	Cane or beet sugar	158	173	21,091	3,164
	including that from Moldova	0	173	17,562	30,382
805	Citrus fruit	16	29	7,343	955
803	Bananas	62	134	16,076	11,575
1006	Rice	120	143	18,252	4,198
	Total				**50,273**

* (Column 2 – Column 1)× Column 3

Domestic Support

Romania's AMS commitment might be a substantial constraint on domestic policy measures. Under the non-accession simulations, in 2006 only sunflower and beef are within the constraints with a protection rate of 0.5 of the published 1998 tariffs, and the same two, plus pork, with a protection rate of 0.25 (table 12.6). Under the accession situation this group also includes poultry and eggs. Expenditure on all other products exceeds the commitments. The accession simulations are interesting in order to see the effects of the adoption of different EU policy instruments. The change of quota from standard to low levels does not substantially affect the results. However, direct payments make a crucial difference. If they are applied to Romania but are not accepted by the WTO as decoupled, then the AMS excess will be around ECU4 billion. Although the Romanian AMS will be aggregated with that of the EU upon accession, the excess in domestic support might create a sensitive area for discussion during the accession negotiations, particularly if the new WTO negotiations on agriculture qualify the direct payments designed under Agenda 2000 as not fully decoupled.

Export Subsidies

Comparisons between the estimated export subsidy expenditures and the WTO commitments are shown in tables 12.7 and 12.8, under non-accession and accession scenarios respectively. At first glance, the comparisons are striking. Romania will have bigger problems meeting its WTO commitments staying outside the EU. In the 2006 non-accession scenario, with 0.5 of published tariffs being reflected in farmers' prices, problems arise for wheat, maize, sunflower, wine, and cheese; and with 0.25, for wheat, maize, and sunflower. Romania appears to be internationally competitive in cereals and sunflower as shown in chapter 8, and it is also a traditional wine producer, with a strategy to increase its outlets on the world market. In order to export these products, the country has to operate domestically at world market prices, because otherwise it will continuously exceed the WTO constraints.

Under the 2006 accession scenario, the situation is better because the domestic (EU Agenda 2000) prices are closer to world market prices. Producers of cereals will operate at world market prices and therefore be competitive without export subsidies. Problems occur for three products only—sugar, wine, and cheese (see table 12.8). The price reform in the wine sector is yet to start, and the Luxembourg Council Meeting in June 1998 extended the price status quo for another year. There are strong opinions in the European Commission about the need for changes in the sugar regime. If reforms of these commodity regimes are pursued, it is likely that these two products would not create a problem upon accession, or would be less of a problem than is projected here.

Table 12.6 Romania, Estimated AMS Overshoots (million ECU)

Non-accession 2006			Accession 2006				
Commodity	Protection Rate 0.5 of 1998 Customs Tariff	Protection Rate 0.25 of 1998 Customs Tariff	Commodity	Unconstrained Output Zero Direct Payments	Standard Quota Zero Direct Payments	Standard Quota Full Direct Payments	Low Quota Full Direct Payments
Wheat	154	83	Wheat	19	19	794	794
Maize	277	179	Maize	75	75	1,158	1,158
Sunflower	−22	−43	Sunflower	−43	−43	80	80
Potatoes	187	79	Potatoes	187	187	187	187
Sugar	25	10	Sugar	243	103	103	95
Tobacco	29	13	Tobacco	41	30	30	28
Wine	1,034	412	Wine	1,207	1,207	1,207	1,207
Milk	272	76	Milk	438	336	336	325
Butter	4	1	Butter	16	16	16	16
Cheese	153	65	Cheese	280	280	280	280
SMP	8	4	SMP	2	2	2	2
Beef	−46	−70	Beef	−50	−50	80	80
Pork	34	−51	Pork	−114	−114	−114	−114
Poultry	30	6	Poultry	−9	−9	−9	−9
Eggs	39	12	Eggs	−11	−11	−11	−11
Total AMS	2,172	774	Total AMS	2,281	2,028	4,140	4,117

Table 12.7 Romania, Export Subsidies, Non-accession 2006

	Protection Rate 0.5 of 1998 Customs Tariff						Protection Rate 0.25 of 1998 Customs Tariff					
Commodity	Price Gap (ECU/t)	Tradable Surplus ('000 t)	Quantity Commitments ('000 t)	Budgetary Outlays (m ECU)	Budgetary Outlay Commitments (m ECU)	Problems	Price Gap (ECU/t)	Tradable Surplus ('000 t)	Quantity Commitments ('000 t)	Budgetary Outlays (m ECU)	Budgetary Outlay Commitments (m ECU)	Problems
Wheat	17	2,654	194	44	15	*	8	2,239	194	18	15	*
Maize	18	2,304	41	42	3	*	9	2,320	41	21	3	*
Sunflower	35	18	1	1	0	*	18	19	1	0	0	*
Sugar	74	109	141	8	17		37	45	141	2	17	
Wine	127	1,252	8	159	2	*	63	–2,973	8	–188	2	
Butter	480	0	14	0	5		240	–2	14	0	5	
Cheese	627	17	11	11	1	*	314	–35	11	–11	1	
SMP	455	2	0	1	0		227	0	0	0	0	
Beef	396	–224	7	0	1		198	–272	7	0	1	
Pork	367	–165	111	0	14		183	–269	111	0	14	
Poultry	342.3	–209	26.5	0	6.7		1,71.2	–256	26.5	0	6.7	

* Problems

Table 12.8 Romania, Export Subsidies, Accession 2006

			Standard Quota			
Commodity	Price Gap (ECU/t)	Tradable Surplus ('000 t)	Quantity Commit-ments ('000 t)	Budgetary Outlays (m ECU)	Budgetary Outlay Commit-ments (m ECU)	Problems
Wheat	0	1,817	194	0	15	
Maize	0	970	41	0	3	
Sunflower	18	−68	1	0	0	
Sugar	338	143	141	48	17	*(outlay)
Wine	141	2,071	8	292	2	*
Butter	1,190	5	14	6	5	
Cheese	980	73	11	72	1	*
SMP	0	3	0	0	0	
Beef	367	−231	7	0	1	
Pork	0	−383	111	0	14	
Poultry	0	−371	27	0	7	

* Problems

Thus, upon amalgamation of the Romanian and EU schedules, Romania might aggravate the EU situation for sugar, wine, and cheese, with the extent also depending on the share of Romanian exports to third countries. On the other hand, with the assumed 3 percent annual increase in real incomes in Romania, by 2006 the country will be in deficit for meats. As Romania will be in the single market, this could slightly relieve the problem that current EU members might have in meeting the export subsidy commitments for meats, particularly beef.

Conclusion

Romania will not create a problem for the EU upon accession in relation to harmonization of tariff bindings. In most cases, the alignment will mean Romania decreasing its bindings. This will not cause political problems for Romania, because the bindings are not operational and such high border tariffs are not needed to protect its domestic farmers. Requests for compensation for decreased market access will be limited to a few products and the amounts are not significant. However, the example of the zero tariff on imports of sugar from Moldova, within the framework of a bilateral free trade agreement, is an illustration of the interaction between existing arrangements and future accession to the EU.

During the pre-accession period, domestic support and export subsidy commitments might create a problem for Romania for several main agricultural products. This problem is not yet visible because of the lack of competitive domestic markets and the resulting lack of transmission of border measures to farmers' prices. However, with the progress of reforms and market developments, the constraints set by domestic support and export subsidy commitments will become tighter. This makes a strong case why pre-accession Romania has to employ policy measures that do not affect farmers' prices, that is, that are decoupled, and must not rely on high tariff protection.

Acknowledgments

We are indebted to Professor Louis Mahé, who launched the study and whose general approach we employ (Mahé 1996), and also to Florica Grigore, who provided technical assistance.

The chapter is adapted from Davidova, S., Giurca, D., Hubbard, L. J., and Rusali, M. A. 1999. WTO Commitments and CP Adoption in Central and East European Countries: The Case of Romania. *MOCT MOST* 9:273–290. Copyright 1999, with permission from Kluwer Academic Publishers.

Notes

1. These measure the degree of protection afforded to domestic producers (see chapter 8).

References

Agra Europe. 1999. No. 1831. January 8.

Buckwell, A., Haynes, J., Davidova, S., Courboin, V., and Kwiecinski, A. 1995. *Feasibility of an Agricultural Strategy of Central and Eastern Europe for*

EU Accession. Final Report to the European Commission, DGI. Mimeo-graphed.

Davidova, S. and Ivanova, I. 1997. Study on Bulgaria and Romania, in the CAP and enlargement: agrifood price developments in five associated countries. *European Economy*. 2:105–170.

De Zeeuw, A. 1997. International agricultural trade negotiations under GATT/WTO: experiences, future challenges and possible outcomes. *European Review of Agricultural Economics* 24:(3–4) 470–479.

Gorton, M. and Deaconescu, D. 1998. *Domestic Resource Costs and International Competitiveness of Romanian Agriculture*. PHARE Project EU Alignment and Policy Reform Report. Mimeo-graphed.

Josling, T. 1996. *Emerging Issues for the 1999 Round: Options for Agricultural Policy Reform*. Paper presented at the Fourth World Bank Conference on Environmentally Sustainable Development, 25–27 September. Washington D.C.

Josling, T., Tangermann, S. and Warley, T. 1996. *Agriculture in the GATT*. Houndmills and London: MacMillan Press Ltd.

Mahé, L. 1996. *The Compatibility of the GATT Commitments of the Czech Republic with those of the EU: Potential Accession Issues*. PHARE Project Czech Republic Policy Advisory Unit: Czech Ministry of Agriculture. Mimeo-graphed.

Organization for Economic Cooperation and Development (OECD). 1994a. The Uruguay Round Agreement on Agriculture: Some Impacts on Central and East European Countries. OECD *Ad Hoc* Group on East/West Economic Relations in Agriculture, AGR/EW/EG(94)26.

——— 1994b. Impact of the Uruguay Round Agreement on Agricultural Commodity Markets: Implications for NIS and CEECs. OECD *Ad Hoc* Group on East/West Economic Relations in Agriculture. AGR/EW/EG(94)28.

——— 1994c. Expected Impacts of the Uruguay Round Agreement on CEEC and NIS Agriculture. OECD *Ad Hoc* Group on East/West Economic Relations in Agriculture. Room document 4.

——— (1995). *The Uruguay Round: A Preliminary Evaluation of the Impacts of the Agreement on Agriculture in the OECD Countries*. Paris: Organization for Economic Cooperation and Development

Plunkett, D. and Maxwell D. 1994. *Future Prospect for Central and Eastern European Agriculture*. USDA, ERS. Working paper.

Romanian Customs Tariff, 1998.

Tangermann, S. 1996. Implementation of the Uruguay Round Agreement on Agriculture: Issues and Prospect. *Journal of Agricultural Economics* 47:(3) 315–337.

Tangermann, S. and Josling, T. 1994. *Pre-Accession Agricultural Policies for Central Europe and the European Union*. Final Report to the CEC, DGI.

Tesliuc, E. 1996. *Regimul Comercial cu Produse Agricole in Romania.* World Bank. Internal paper. Mimeo-graphed.
Uruguay Round (UR) Schedule of Romania. 1996. Geneva, Switzerland: World Trade Organization.
Uruguay Round (UR) Schedule of the EU15. 1996. Geneva, Switzerland: World Trade Organization.

Chapter 13

Conclusion: Knowns, Unknowns, and Possibilities

Kenneth J. Thomson

The preceding chapters of this book have reviewed a number of aspects of Romanian agriculture and rural areas, with a view to the eventually successful accession of Romania to the EU. By the time they are read, some of their data and/or conclusions may be out of date, as is inevitable in a book that comments on a country and economy "in transition." Unsurprisingly, the contents are also far from comprehensive. It seemed that other aspects of Romanian agriculture, especially perhaps those of a more noneconomic nature—such as legal, technical, or institutional analyses—were, although important, likely to widen the scope of the book too far.

The temptation to compare Romania with other EU applicant countries in Central and Southeastern Europe has been largely resisted in this book. Of course, every country has its own historical individualities, and, geographically, Romania is the furthest applicant country, other than Bulgaria and Turkey, from the main EU-15 centers. Indeed, Romania has its own very special characteristics of political culture: "Communism had the paradoxical effect of separating Romania from the West without ultimately bringing it closer to Russia" (Boia 2001, 205).

Thus, it was considered most profitable to analyze Romania in its own terms, with the EU context firmly in mind, rather than exploring perhaps misleading similarities and contrasts with other CEECs.

This final chapter comprises first a selective summary of the book's contents, picking out those conclusions that seem most relevant to an overall economic assessment of Romanian agriculture.

After this, some gaps and difficulties in the current situation are identified as deserving particular attention, followed finally by some wider considerations related to Romania's entry into the EU and its adoption of the EU's agricultural and rural policies.

A Selective Summary

The initial background chapters (1 and 2) describe the hesitant and partial approach to economic reform taken by Romania over the decade of the 1990s. As with other sectors, agriculture and food processing had been very firmly under state control for several decades. After 1989, several million "independent" semisubsistence farm households were created more or less immediately, and have co-existed alongside the remains of state farms mostly transformed into commercial companies (the CCSCs), which often enjoyed substantial state support if only via debt forgiveness. The first-stage downstream food-processing sector was left largely unchanged for a long period, though with less reliable farm supplies and retailing customers. Uncertainty persisted as to whether government would take further steps to liberalize and de-monopolize the processing sector, or would be able to reassert its leading role, and this has led to a major degree of industry stagnation.

Alongside these economic uncertainties was the fitful progress toward land reform, in terms of not only the necessary legislative measures but also the implementation of land sales and leasing by the agencies responsible. In turn, this explains the slow (almost negligible) improvement in credit conditions for Romanian agriculture and agri-food businesses, although the rather chaotic reconstruction of the entire finance sector has also been to blame.

As the following chapters (3, 4 and 5) make clear, improved farm credit conditions are essential if individual producers are to prosper in the new Romania. Without access to capital in the form of short-term loans for seed, fuel, and other variable inputs, and to the services (if not ownership) of machines, stores, irrigation, etc., it cannot be expected that such farmers can better their position. The government's voucher schemes of 1998–2000 addressed the first of these problems (see chapter 6), but was expensive to the state budget, and ran counter to both the EU and IMF approaches to public finance. Larger-scale and longer-term credit will depend on the creation of lending facilities—private or public (or a mixture of the two)—targeted at agriculture, which will otherwise find it hard to compete with more attractive options for domestic and foreign investors. But such targeting runs several risks; for example, "leakage" to other sectors, inefficient investment, and irrecoverable defaults. However, credit is not the sole, or even main, prerequisite for agricultural recovery in Romania. As well as the necessary completion of land market reforms, and a generally more stable macroeconomic environment, farmers need more "transparency" and stability in marketing conditions, better infrastructure in terms of roads and telecommunications, and improved education and training.

Even with all this, it may be doubted if farming in Romania can offer a sufficient number of small farming households a satisfactory standard of living, let alone form the basis for successful commercial family farming. Chapters 7 and 8 show that much recent "agricultural policy" in Romania, such as attempts at credit provision and product price stabilization, has in fact had only limited effect on the many semisubsistence holdings, which are distanced from commodity markets by a combination of low marketed shares of production and inefficient trading relations with processing enterprises. Chapter 8 explores the fundamental question of Romania's comparative advantage in agricultural production, and concludes that wheat, maize, and sunflower offer greatest promise, while sugar beet, and the ruminant sectors of dairying and beef production, appear less attractive. The intensive sectors of poultry and pork also fare less well in potential economic performance at world (1998) prices, but efficient modern plant and management might reverse this conclusion.

Of course, comparative advantage has to be judged against the actual or likely trading conditions, and these include not only the normal fluctuations of the world markets, but also uncertainties over the import and export behavior of the main nearby trade blocs of the EU and Russia and her ex-satellites. EU accession will involve adoption of a CAP currently facing renewed pressure for further reform. However, it may be doubted if the EU will move too far from its recent pattern of only limited protection to most crops—with the notable exception of sugar—and to the intensive pig and poultry sectors, while maintaining substantial support for dairy, beef, and other grazing livestock. At any rate, as chapter 9 shows, the

Europe Agreement with the EU over farm trade has had only limited effect to date.

One strategic option, explored in some detail in chapter 10 but mentioned in several other places in this book, is to focus on the more general development of the rural economy, including farm diversification. Certainly, this is in line with current EU thinking, and offers hope of bypassing some of the well-known struggles to achieve agricultural success. Moreover, with pre-accession SAPARD funding in more immediate prospect than post-accession CAP measures, this route is being explored by Romanian and EU administrations in some detail, albeit still with much emphasis on farm-based and food-related investment and training.

In chapter 11, a simple "model" is used to estimate a number of welfare or real-income effects of adopting the CAP in a hypothetical accession year, 2006. The calculations highlight the importance of focusing on Romanian food consumers as well as farm producers, and put an order of magnitude on state budget effects, in or out of the EU. Not unexpectedly, the question of direct CAP payments to farmers tends to dominate, but the relatively minor significance of, for example, CAP marketing quota allocation is also of interest. Finally, chapter 12 assesses the constraints likely to arise from WTO commitments as Romania approaches and embraces EU accession. The analysis identifies some potential risks, but avoidance of these may help to smooth the path to eventual membership, by preventing the adoption—in better fiscal conditions—of overgenerous internal and border subsidies to Romanian agriculture during the pre-accession period.

Gaps and Difficulties

The above summary suggests some obvious gaps in the coverage of this book. Some can perhaps be filled by more recently acquired knowledge; others will require fresh investigations and analysis. In several cases, it is difficult to go much beyond speculation, since only real-life experience will show what is possible, and what will ensue.

Land tenure, and more generally the overall *farming structure*, in Romania certainly cries out for further consideration. While arguments exist over the exact significance of land ownership as regards farming productivity (Mathijs and Swinnen 1988; Hughes

2000, 61–87), few would maintain the current situation of uncertainty is conducive to a confident and dynamic agricultural contribution to the national economy. Admittedly, it is hard to ask a population which has only recently seen the return of a basic natural asset from a distant and arbitrary state into its own, largely local, hands to place much confidence in a legal system which has yet to exhibit much efficiency or, in some cases, honesty and independence. Yet, without effective and transparent control over landownership, strong farming businesses cannot be built up, on either owned or leased land. The role of land as collateral for borrowing is also crucial in reducing the undercapitalization of Romanian agriculture. Both are inherently long term in nature, as matters of trust and confidence, which underpin the "social economy" underlying the more formal (and informal) contractual arrangements which agriculture requires.

In this book, as elsewhere in writings on Central Europe (Swinnen, Buckwell, and Mathijs 1997), much attention has been paid to the *individual farmer*, whether small-scale (semisubsistence) peasantry or somewhat larger commercially oriented "family farming." These forms have been, or indeed are, the bedrock of West European and much North American agriculture. Yet, family farming has been under pressure from "agribusiness" in the West and elsewhere. The latter appears to be a more efficient attractant and user of resources than the individualized model, especially in conditions of an advancing economy with modern consumer demands, whether for highly processed foods, and/or certificated "organic" products. However, Romania comes from a different structural position, with experience (if not always a happy one) and some legacy (despite depreciation and asset stripping) of much *larger-scale operations*, which may well be more suited to extensive arable enterprises, perhaps combined with feed lots for meat and egg production. More attention should perhaps be paid to revitalizing at least some of the pre-1989 farming structure in Romania, this time on a fully privatized basis. It seems likely that countries such as Hungary will soon show what is possible and profitable in post-transition Central European agriculture, and thereby provide models for Romania and others.

However, the manifest and manifold problems of the large number of *small farm households* cannot be ignored. It is widely agreed (Kostov and Lingard 2002) that these have provided a vital "safety net" for the many (majority) left disadvantaged by the first decade of transition. While their agricultural activities are important in providing food for most of the rural population, it is an open question whether they can really provide the basis for a nationwide farming renaissance. It is more likely that they will continue to provide a relatively secure way of life—including housing and social networks—for those without other means of support. Perhaps the most that the Romanian government can be expected to achieve as far as these holdings and households are concerned is to maintain and adjust welfare payments to ensure a basic income, to encourage those able to lift themselves out of semisubsistence to do so (e.g., via education, training, and marketing assistance), and to prevent their exploitation by more successful persons and businesses. Meanwhile, true "agricultural policy" should focus on large-scale enterprises as suggested above (primarily through institutional instruments such as export marketing initiatives, staff training, public infrastructure) and on encouraging via similar but distinct channels the growth of western-style "family farms," which have access to the necessary expertise and capital to develop new markets, most likely for specialized products, crops, or livestock.

Among these new products, more attention could be paid to commodities not usually studied deeply in broad analyses of agriculture in Romania or other countries. Among these products may be suggested *fruits* and *vegetables*, along with their processed derivatives such as *wine*, fruit juices, and highly packaged catering and home products. A further possibility, especially if the world energy market should turn difficult, would be *biomass and bioenergy crops*, where Romania might have advantages over other countries with less fluid farming systems. In most such cases, detailed commodity and market (demand) knowledge is required, both for analysis and enterprise development. More generally, *product innovation* and *marketing efficiency* are as important to successful agri-food development as farm-based productivity, and need indirect state support via research, development, and marketing aids.

Policy Outlook and Wider Considerations

Since the conception of this book at about the time when the EU enlargement process was being formulated in some detail, the policy context has moved on, though not dramatically. The timetable for agreement by end-2002 between the Common Positions of the EU-15 and the counterproposals of ten out of the twelve current applicants is tight but may indeed be achievable, and would allow some first accessions in 2004. This epochal event will set the tone for future accessions. However, the inclusion of new members from Central Europe will radically alter the character of the Council of Agricultural Ministers and the overarching European Council. While probably reinforcing agricultural protectionism, it may also set up new tensions for further accessions.

More relevantly to the central concerns of this book are the prospects for the CAP itself, currently facing a Mid-Term Review of the 1999 Agenda 2000 decisions but unlikely to undergo major reform before 2008, that is, after the first of the new accessions, and the (possible) full implementation of the Agenda 2000 decisions. For Romania, these uncertainties may be less important than for other Central European countries. EU market support for cereals is already substantially reduced (though depreciation of the dollar could reverse this), and Romania's cattle and sheep sectors are not of major significance. While Romania might have at one time hoped to have benefited from export subsidies for its cereals and from internal CAP disposal mechanisms for occasional surpluses of cereals and wine, the scope of these measures has been severely reduced by both internal CAP reform and the limitations of the EU's WTO commitments, which may be intensified in a future post-Doha agreement.

The historical peculiarities include both the legacy of the Ceaucescu regime in terms of a crudely developed large-farming structure that collapsed after 1989, and also the very uncertain start to the transition process toward market orientation. These and other features have meant that the progress of the eastward EU enlargement process during the 1990s has left Romania far behind most other Central European countries. Even Bulgaria, as the obvious and neighboring comparator, has stolen a march on Romania with its early achievement of a SAPARD funding system. It seemed appropriate, therefore, to consider Romania simply in its own terms, and to leave externally comparative judgments to others.

From an economic point of view, the basic challenge is to identify ways of pursuing comparative advantages—inside or outside the CAP—to the benefit of the country as a whole. The phrase "as a whole" is important. For one thing, it draws attention to the distinction between the public and private sectors, and avoids the still common tendency to equate the interests of the state executive with those of citizens in general. For another, it directs attention at the many competing claims for policy preference, which may be conferred on a number of rational grounds, including the correction of market failures and the reduction of social inequalities, as well as less defensible ones such as the maintenance of privilege. Given the large number of claims on scarce public resources in Romania, and natural impatience to overcome past difficulties, choices will be difficult, particularly in the context of inadequate and uncertain information.

Meanwhile, Romania has not been forgotten by the international community. Significant EU assistance is available in the form of its pre-accession instruments, including SAPARD, for which Romania will be one of the biggest recipients. Both the European Investment Bank and the World Bank are supporting projects relating to agricultural and rural development in Romania. The EU Commissioner for Enlargement, Günter Verheugen, has recently said:

> For Bulgaria and Romania, the objective is to open negotiations on all chapters in 2002. Moreover, the Commission will propose for all countries with which negotiations cannot be concluded by the end of this year, an enhanced pre-accession strategy and a precise road map with the objective to further speed up the negotiations. My message to you Mr. President Iliescu is the same, which I passed to the Bulgarian government some days ago: Romania and Bulgaria will not be forgotten or left behind.
> (Speech to Forum Europe conference *EU Enlargement: The Key Questions*, Brussels, 28 April 2002)

This gives some reassurance that the apparent success of some Central European countries in progressing further with accession negotiations will not disadvantage Romania, which may indeed benefit from the undoubted difficulties that EU entry will bring.

Romania's location on the Black Sea at the southeast corner of an expanded EU, and in control of the Danube once that waterway resumes its traditional role as a continental transport corridor, suggests that the country could play a major role in trade with countries further afield, almost regardless of EU accession. Together with its assets in terms of large areas of excellent agricultural soils and relatively good major roads, the agricultural potential of the country—especially if economic prosperity can be achieved around the Black Sea as well as domestically—is obvious. The basic question is whether the political and managerial elite of Romania can rise to this challenge.

References

Boia, L. 2001. *Romania.* London: Reaktion Books.

Hughes, G. 2000. Total productivity of emergent farm structures in Central and Eastern Europe. In *Central and Eastern European Agriculture in an Expanding European Union*, ed M. Banse and S. Tangermann, 61–87. Wallingford: CABI Publishing.

Mathijs, E. and Swinnen, J. 1988. The economics of agricultural decollectivization in East Central Europe and the former Soviet Union. *Economic Development and Cultural Change* 47(1):1–26.

Swinnen, J., Buckwell, A., and Mathijs, E. (eds.) 1997. *Agricultural Privatisation, Land Reform and Farm Restructuring in Central and Eastern Europe.* Aldershot, UK: Ashgate.

Kostov, P. and Lingard, J. 2002. Subsistence farming in transitional economies: lessons from Bulgaria. *Journal of Rural Studies* 18 (1):83–94.

Index

About the Contributors

Dr. Gertrud Buchenrieder (née Schrieder) is Assistant Professor, University of Hohenheim, Germany.

Ms. Adriana Cristoiu is Extension Service Officer, ANCA, Ministry of Food and Agriculture, Bucharest, Romania, and Ph.D. student, Imperial College Wye, University of London, UK.

Dr. Christine Coulomb is Financial Asset Manager, Paris, France.

Dr. Graham Dalton is Honorary Senior Lecturer, University of Aberdeen, UK.

Dr. Sophia Davidova is Senior Lecturer, Imperial College Wye, University of London, UK.

Dr. Junior Davis is Economic Policy Adviser, Department for International Development, UK.

Mr. Dinu Deaconescu is former Counterpart Officer for the EU-PHARE "EU Alignment and Policy Reform," Ministry of Food and Agriculture, Bucharest, Romania.

Dr. Marc Duponcel is Policy Officer, subregional office for Central and Eastern Europe, FAO.

Dr. Daniela Giurca is Scientific Secretary of the Institute of Agricultural Economics, Bucharest, Romania.

Dr. Matthew Gorton is Lecturer, University of Newcastle, UK.

Mr. Paul Hill is Senior Lecturer, Imperial College Wye, University of London, UK.

Dr. Lionel Hubbard is Senior Lecturer, University of Newcastle, UK.

Dr. John Lingard is Senior Lecturer, University of Newcastle, UK.

Dr. Dan Petrovici is Lecturer, University of Canterbury at Kent, UK.

Dr. Marian Rizov is at LICOS Centre for Transition Economics, Department of Economics, Katholieke Universiteit Leuven, Belgium.

Ms. Mirela Rusali is Research Fellow, Institute of Agricultural Economics, Bucharest, Romania.

Dr. Bill Slee is Senior Lecturer, University of Aberdeen, UK.

Professor Johan Swinnen is Professor of Development Economics and Food Policy, Department of Economics, Katholieke Universiteit Leuven, Belgium.

Professor Kenneth Thomson is Professor of Agricultural Economics, University of Aberdeen, UK.

Professor Paul Webster is Professor of Agricultural Business Management, Imperial College Wye, University of London, UK.

Mr. Nigel Williams is Senior Lecturer, Imperial College Wye, University of London, UK.

www.ingramcontent.com/pod-product-compliance
Ingram Content Group UK Ltd.
Pitfield, Milton Keynes, MK11 3LW, UK
UKHW041632031026
15587UKWH00001B/222

9 780739 105184